What Is Life ?

1
Earth in space.

Between these two realms

Mycoplasma. Phylum:
Aphragmabacteria.
Kingdom: Monera. Some
of the smallest bacteria
(magnified 10,000 times).

lies all of life as we know it.

What Is Life?

Lynn Margulis and Dorion Sagan

Foreword by Niles Eldredge

A Peter N. Nevraumont Book

SIMON & SCHUSTER

New York London Toronto Sydney Tokyo Singapore

SIMON & SCHUSTER
Rockefeller Center
1230 Avenue of the Americas
New York, New York 10020

10 9 8 7 6 5 4 3 2 1

Printed in Italy

Library of Congress Cataloging-in-Publication Data

Margulis, Lynn, 1938-
    What is life? / Lynn Margulis and Dorion Sagan :
    introduction by Niles Eldredge.
        p. cm.
    "A Peter N. Nevraumont book."
    Includes bibliographical references.
    ISBN 0-684-81326-2
    1. Life (Biology) 2. Biology—Philosophy.
3. Biological diversity. 4. Life—Origin.
I. Sagan, Dorion, 1959- .II. Title.
QH326.M298 1995
577—dc20

Contents

Foreword
Undreamt Philosophies

Why has evolution crafted a sentient species? Why did our consciousness, our realization of our very existence, evolve? What purpose does it serve? I am persuaded by behaviorist Nicholas Humphries' conjecture that, in being able to consult their inner selves, our ancestors gained insight on the minds of their mates, offspring, and other members of their social bands. Knowing thyself is the best way to knowing others, and thus an advantage in negotiating the complexities of daily social life.

We humans are, of course, animals. I have long thought that the very best insight into what it means to be a living, breathing animal is simply to consider one's very own life. However far our cognitive, cultural capacities have taken us from traditional existence within local ecosystems, we nonetheless still obtain energy and food to develop, grow and maintain our corporeal existence. Many of us (perhaps too many of us) also engage in reproduction. As Lynn Margulis and Dorion Sagan tell us in **What Is Life?**, the business of maintaining corporeal existence and reproducing are quintessential activities, the very hallmarks of life. To know oneself as an organism, then, is to establish quite a few of the very basics of all living systems.

But humans, of course, do not constitute the entire biological universe. We are but one species of tens of millions now inhabiting planet Earth. And so we cannot expect to divine all of life's mysteries, all the different nuances of what it means to be alive, simply by consulting our inner selves. There are inherent limits to the revelatory principle of knowing thyself in order to know the world. But even I, a seasoned practitioner in evolutionary biology, was not fully prepared for the wild spectrum of life presented to us by Lynn Margulis and Dorion Sagan in **What Is Life?**. For in these pages we meet organisms vastly different from ourselves. And we encounter ways of thinking about life that could not possibly arise from simple introspection.

What is Life? is a feast of biological and intellectual diversity. Here we meet microbes—microscopic organisms—for which oxygen is a poison, and others who "breathe" sulfur compounds. And still others which feed on hydrogen and carbon dioxide using neither energy from the sunlight, nor that from the flesh of others. We encounter bacteria routinely exchanging genetic materials with other species—even after billions of years of evolutionary separation. We see the entire outer rind of Earth portrayed in convincing fashion as a single, mega-living system. And we learn that the evolutionary process that has produced this prodigious array has done so in astonishing ways—melding separate, simple organisms more than once to produce more complex descendant species. And therein lies a particularly interesting saga of intellectual sleuthing and derring do.

Darwin taught us that all of life is descended from a single common ancestor. In *What Is Life?*, Margulis and Sagan tell us the amazing fact that not only are our own mammalian, nucleated ("eukaryotic") cells descended from ancient bacteria, **they are literally amalgams of several different strains of bacteria.** Amazing! Stranger than fiction! And undreamt of in traditional biological philosophies—until Lynn Margulis began her research a quarter of a century ago.

Lynn Margulis has achieved what every scientist dreams of, but few are destined to accomplish: she has rewritten the basic textbooks. She conceived of a logical, yet audacious explanation of an outstanding fact. Human cells, like those of all animals, the eucalyptus tree and the mushroom, have most, but not all, of their DNA corralled into a cellular nucleus, neatly walled off from the various organelles that dot the plains of their typical cell's cytoplasm. It was the "not all" that attracted her attention: some of these extra-nuclear organelles—specifically, the power plants of all animal and plant cells, the "mitochondria"—were also known to have their own DNA. In plants, both mitochondria and chloroplasts, the locus of photosynthesis, have their own DNA complements. The simple question she faced was: why? Why is there an independent set of genes in these cytoplasmic organelles, when all of the "normal" genetic material is otherwise organized as double sets of chromosomes within the bounds of the nuclear walls?

Biological structures are signals of ancient evolutionary events. We owe the five fingers on our hands not to novel evolutionary events a million years ago on the African savannahs, but rather as a holdover from the original complement of five digits on the forefoot of the earliest land vertebrates ("tetrapods"), who evolved some 370 million years ago.

So, too, is mitochondrial DNA a holdover, a signal of an evolutionary event. But this was like no other event ever proposed in evolutionary annals: Lynn Margulis, to her everlasting credit, saw that separate DNA complements **imply the fusion of at least two different kinds of other organisms, each with its own DNA complement, to form a single, complex "eukaryotic" cell.** Initially condemned as heresy, this elegant idea had so much going for it that the biological world has long since accepted it. There is simply no other plausible explanation for the existence of separate DNA complements in a "single" cell.

In *What Is Life?*, Lynn Margulis and Dorion Sagan tell us precisely which kinds of bacteria fused to form the original nucleated cells—*our* cells. But that is far from all, for the Margulis mind, ever restless, has kept on pushing the envelope. *What Is Life?* presents the case for an even **earlier** evolutionary fusion of bacteria species. Margulis has come to be convinced that such symbiotic origins of novel life forms ("**symbiogenesis**") has been far more common than ever dreamt by evolutionary biologists steeped in the Darwinian tradition—a tradition that emphasizes competition far more than cooperation in the evolutionary process. Symbiogenesis is

Margulis' central contribution to the evolutionary dialogue, which has become enriched through her efforts to see the grand implications latent in the history of the microbial world.

But there is more to the Margulis-Sagan canon than even these profoundly new, and heretofore undreamt, philosophies. Tireless champions of the microbial world, the authors have labored mightily in an almost public-relations sense, striving to reveal the immensely diverse array of micro-organisms. For microbes will not only inherit the earth (should, for example, we complex multicellular creatures fall prey to the next spasm of mass extinction); for microbes got here long before we did, and in a very real sense they already "**own**," and most certainly **run**, the global system. They fix and recycle nitrogen and carbon and other essential elements otherwise unavailable to our bodies; they produce oxygen, natural gas (methane), and so on and on. Without the microbial world, life as we ourselves experience it simply could not be.

All of which lifts the Margulis gaze from the microscopic to the global: Earth truly is a living system, a globally pulsing amalgam of organisms and the physical "inanimate" world. Whether or not one chooses to call this system "Gaia" and pronounce it as alive as any organism does not, in a profound sense, really matter. For in reading *What Is Life?*, we see, clearly and simply, that the global system linking life with the physical realm truly does exist, and that we humans, despite appearances and protestations to the contrary, are still very much a part of that system.

Which takes us back to the ultimate value of being aware of our own existence. As we read *What Is Life?*, we think about life's riotous diversity and evolution's exuberance, and we realize that the global system, all that life, and, in the end, *our* very own existence, are very much under threat—from our very own selves. *What Is Life?* combines the stranger-than-fiction realities of the living world

with the kind of intellectual force that can reveal new undreamt philosophies. It yields the understanding we so desperately need if we are to confront the mounting threat we humans pose to the global ecosystem as we cross over the Millennial divide. Knowledge is power, and *What Is Life?* equips us with an understanding of the living world that we so desperately need if we—along with the world's ecosystems—are to survive.

Niles Eldredge

*American Museum of
Natural History*

1

Life: The Eternal Enigma

Life is something edible, lovable, or lethal.

—**James E. Lovelock,** inventor

Life is not a thing or a fluid any more than heat is. What we observe are some unusual sets of objects separated from the rest of the world by certain peculiar properties such as growth, reproduction, and special ways of handling energy. These objects we elect to call "living things."

—**Robert Morison,** physicist

In the Spirit of Schrödinger

Half a century ago, before the discovery of DNA, the Austrian physicist and philosopher Erwin Schrödinger inspired a generation of scientists by rephrasing for them the timeless philosophical question: *What Is Life?* [PLATE 3] In his classic 1944 book, bearing that title, Schrödinger argued that, despite our "obvious inability" to define it, life would eventually be accounted for by physics and chemistry. Life, Schrödinger held, is matter which, like a crystal—a strange, "aperiodic crystal"— repeats its structure as it grows. But life is far more fascinating and unpredictable than any crystallizing mineral:

> The difference in structure is of the same kind as that between an ordinary wallpaper in which the same pattern is repeated again and again in regular periodicity and a masterpiece of embroidery, say a Raphael tapestry, which shows no dull repetition, but an elaborate, coherent, meaningful design traced by the great master.[1]

Schrödinger, a Nobel laureate, revered life in all its marvelous complexity. Indeed, although he devised the wave equation that helped give quantum mechanics theory a firm mathematical basis, he never conceived of life as simply a mechanical phenomenon.

Our book, addressing life's fulness without sacrificing any science, reproduces not only Schrödinger's title but also, we hope, his spirit. We have tried to put the life back into biology.

What is life? is surely one of the oldest questions. We live. We—people, birds, flowering plants, even algae glowing in the ocean at night—differ from steel, rocks, inanimate matter.

We are alive. But what does it mean to live, to be alive, to be a discrete being at once part of the universe but separated from it by our skin? What is life?

3

Erwin Shrödinger: a physicist whose emphasis on the physiochemical nature of life helped inspire the discovery of DNA and the molecular biological revolution.

Thomas Mann gave an admirable, if literary, answer in the novel *The Magic Mountain*:

> What was life? No one knew. It was undoubtedly aware of itself, so soon as it was life; but it did not know what it was ... it was not matter and it was not spirit, but something between the two, a phenomenon conveyed by matter, like the rainbow on the waterfall, and like the flame. Yet why not material—it was sentient to the point of desire and disgust, the shamelessness of matter become sensible of itself, the incontinent form of being. It was a secret and ardent stirring in the frozen chastity of the universal; it was a stolen and voluptuous impurity of sucking and secreting; an exhalation of carbonic gas and material impurities of mysterious origin and composition.[2]

Our ancestors found spirits and gods everywhere, animating all of nature. Not only were the trees alive but so was the wind howling across the savanna. Plato, in his dialogue *Laws*, said that those perfect beings, the planets, travel around the earth voluntarily in circles. Medieval Europeans believed the microcosm, the small world of the person, mirrored the macrocosm, the universe; both were part matter and part spirit. This ancient view lingers in the animals of the zodiac and in the astrological notion that celestial bodies influence mundane ones.

In the seventeenth century the German astrologer-astronomer Johannes Kepler (1571-1630) calculated that planets travel around the sun in ellipses. Nevertheless, Kepler (who wrote the first work of science fiction and whose mother was arrested as a witch) believed that the stars inhabit a three-kilometer-thick shell far beyond the solar system. He considered Earth a breathing, remembering, habit-forming monster. Although Kepler's view of a living Earth now seems whimsical, he reminds us that science is asymptotic: it never arrives at but only approaches the tantalizing goal of final knowledge. Astrology gives way to astronomy; alchemy evolves into chemistry. The science of one age becomes the mythology of the next. How will future thinkers assess our own ideas? This movement of thought—of living beings questioning themselves and their surroundings—is at the heart of the ancient question of what it means to be alive.

Life—from bacterium to biosphere—maintains by making more of itself. We focus on self-maintenance in our first chapter. Next, we trace views of life from very early on through European mind-body dualism and then to modern scientific materialism. Chapter 3 explores life's origins and its memory-like preservation of the past. Our ancestors—the bacteria that brought Earth's surface to life—are featured in chapter 4.

Through symbiotic mergers, bacteria evolved into the protists of chapter 5. Protists are unicells, including algae, amebas, ciliates, and other postbacterial cells with erotic habits anticipating our own; they evolved into multicelled beings experiencing sex and death. We call the unicellular protists, together with their close multicellular relatives — some of which are very large — protoctists. The bacteria that formed protoctists were to have a spectacular future. They became animals (chapter 6), fungi (chapter 7), and plants (chapter 8). In the last chapter we pursue the unorthodox but common-sensical idea that life — not just human life but all life — is free to act and has played an unexpectedly large part in its own evolution.

Life's Body

Life, although material, is inextricable from the behavior of the living. Defying definition — a word that means "to fix or mark the limits of" — living cells move and expand incessantly. They overgrow their boundaries; one becomes two become many. Although exchanging a great variety of materials and communicating a huge quantity of information, all living beings ultimately share a common past.

Perhaps even more than Schrödinger's "aperiodic crystal," life resembles a fractal — a design repeated at larger or smaller scales. [PLATE 4] Fractals, beautiful for their delicacy and surprising in their apparent complexity, are produced by computers, as graphics programs iterate, or repeat, a single mathematical operation thousands of times. The "fractals" of life are cells, arrangements of cells, many-celled organisms, communities of organisms, and ecosystems of communities. Repeated millions of times over thousands of millions of years, the processes of life have led to the wonderful, three-dimensional patterns seen in organisms, hives, cities, and planetary life as a whole.

Life's body is a veneer of growing and self-interacting matter encasing Earth. Twenty kilometers thick, its top is the atmosphere and its bottom is continental rock and ocean depths. Life's body is like a tree trunk. Only its outermost tissues grow. Unless protected by technology, itself an extension of life, any individual removed from the living sphere is doomed.

Life, as far as is known, is limited to the surface of this third planet from the sun. Moreover, living matter utterly depends on this sun, a medium-sized star in the outback of the Milky Way Galaxy. Less than one percent of the solar energy that strikes the earth is diverted to living processes. But what life does with that one percent is astounding. Fabricating genes and offspring from water, solar energy, and air, festive yet dangerous forms mingle and diverge, transform and pollute, slaughter and nurture, threaten and overcome. Meanwhile, the biosphere itself, subtly changing with the comings and goings of individual species, lives on as it has for more than three thousand million years.

Animism vs. Mechanism

If you wish, you can reach for a glass of water or snap this book shut. From the experience of willing our bodies to move came animism: the view that winds come and go, rivers flow, and celestial bodies guard the heavens because something inside each wills the movement. In animism all things, not just animals, are seen to be inhabited by an inner, animating spirit. Formalized in polytheistic religion, the multiplicity of gods — a moon god, earth god, sun god, wind god, and so on — was replaced in Islam, Judaism, and Christianity by a single god who crafted the world. Winds and rivers and celestial bodies lost their will, but living organisms — especially humans — retained theirs.

Finally, the last outposts of animism — living organisms — yielded to the philosophy of mechanism. Motion need not imply any inner consciousness; the program could have been "built in" by a creator. Wind-up toys and automated models of the solar system suggested to their inventors that even living things may be constructible from lifeless mechanisms, subtle concealed springs, tiny unseen pulleys, levers, cogs, and gears. Comparing flowing

Like Julia and Mandelbrot patterns, Peter Allport's duality fractal exemplifies a crucial trait of life: the iterative production of complex entities from a component whose design repeats at ever greater scales. Whereas computer graphics are due to the repetition of an algorithm, the motive force behind "living fractals" is the reproduction of cells.

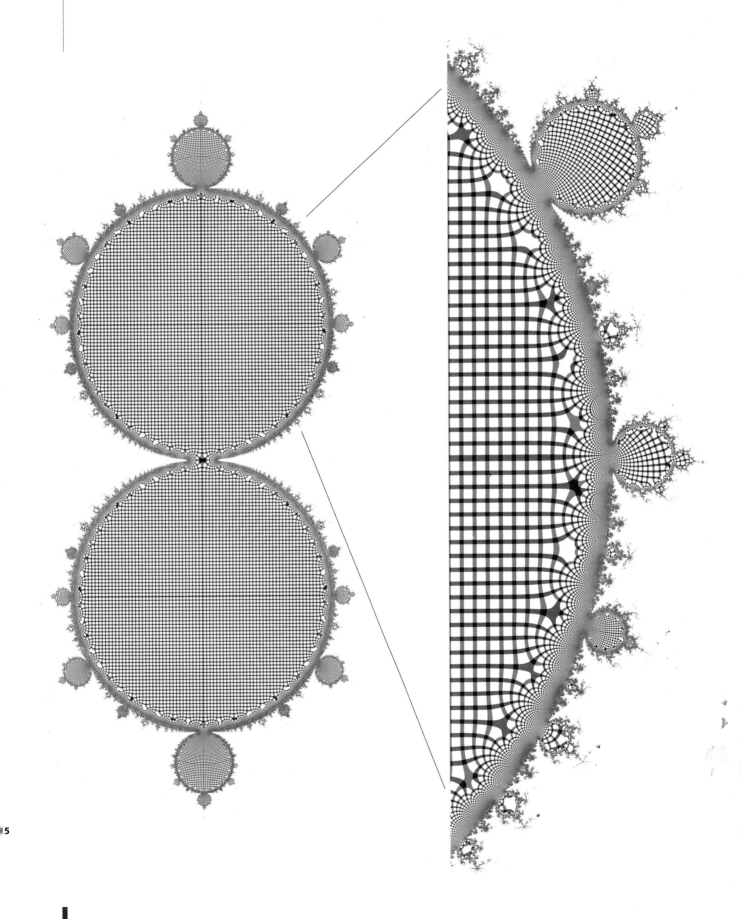

blood to a hydraulic system, the heart to a pump, English physician William Harvey (1578-1647) discovered circulation of the blood. Scientists sleuthed out the world's secret mechanisms, part of an overall design. Natural History revealed the world to be a giant mechanism made according to the mind of God.

Isaac Newton (1642-1727) became the high priest of mechanism. A devoted student of alchemy, scripture, and the occult, Newton made unparalleled innovations in optics, physics, and mathematics. In doing so he helped bridge the gap from the medieval cosmos to the modern one. Explaining the motions of the planets with a new law of gravity, Newton's equations showed that the world of the heavens and that of Earth were one and the same; the force that kept the moon in orbit was also the force that thuds an apple to the ground. So revealing were Newton's discoveries of "laws" governing the entire universe that to some it seemed he had—in Kepler's words—"glimpsed the mind of God." Inspired by Newton's analyses, (Pierre-Simon) marquis de Laplace (1749-1827) speculated that, with sufficient information, the entire future of the universe, even the most minute human action, could be predicted. Far from being moved by hidden spirits, the celestial bodies now seemed to be under the governance of preexistent mathematical laws. Divine intervention became increasingly superfluous. God did not need to fiddle with creation. He had crafted it to last. The cosmos worked itself.

With a grasp of gravitation's cosmic sweep, scientists were spurred to explore phenomena once considered beyond the ken of human comprehension. Electricity and magnetism, chemicals and colors, radiation and heat, explosion and chemical change were all described with an eye to their underlying unity. Optical instruments, telescope and microscope, presented formerly unseen worlds of the very far and the very near. Experiment and criticism replaced blind acceptance of classical authority and divinely revealed truth. Scientists coaxed nature to yield some of her most private secrets. Oxygen's role in fire, lightning as electrical discharge, gravity as the invisible force causing the tides and attracting the moon into Earth's orbit—one by one nature laid down her cards.

Under the spell of the mechanical world view, the ancient alchemical dream of shaping nature to human will became technological reality. After centuries of meddling with steamy concoctions in a Faustian quest to be godlike, a 1953 discovery seemed to reveal the very secret of life. Life was chemical and the material basis of heredity was DNA, whose helical and staircase-like structure made clear how molecules copied themselves. Indeed, the "aperiodic crystal" that Schrödinger had predicted was uncannily similar to the double helix first described by the English chemist Francis Crick and American whiz kid James D. Watson. Replication was no longer beholden to a mysterious "vital principle"; it was the straightforward result of interacting molecules. The description of how DNA fabricated a copy of itself out of ordinary carbon, nitrogen, and phosphorus atoms was perhaps the most spectacular of all mechanism's successes. But paradoxically, this success born of self-directed minds seemed to portray life—including the scientists themselves—as the result of atoms involuntarily interacting according to changeless and inviolable chemical law.

Between these two extremes—the entire universe as alive, and the living organism as chemical and physical machine—lies the panorama of opinion. But is there not something wrong with both the mechanization of life *and* the vitalization of matter?

The world as a vast machine fails to account for our own self-awareness and self-determination because the mechanical worldview denies choice. Mechanisms, after all, don't act; they react. And mechanisms, moreover, don't come into existence on their own. The assumption that the universe is a mechanism implies that it was made according to some humanlike design—that is, by some living creator. In other words, successful as it is, the scientific mechanistic worldview is deeply metaphysical; it is rooted in religious assumptions.

The animistic view of the cosmos as a huge organism is also flawed. It blurs the distinctions among what is living, what is dead, and what has never been alive. If everything were alive, there would be no interest in—and scientists never would have discovered the replicative chemistry of—life.

We thus reject mechanism as naive and animism as unscientific. Even so, life, as an emergent behavior of matter and energy, is best known by science. Schrödinger was correct in advocating a search for the physico-chemical underpinnings of life. So are Watson and Crick and other physicists and molecular biologists who hail the structure of DNA as a key to life's secrets. Like an uncoiling spring pushing the soft gears of life, DNA copies itself as it directs the making of proteins that together form the leopard's spots, the spruce tree's cone, and living bodies in general. Understanding how DNA works may be the greatest scientific breakthrough in history. Nonetheless, neither DNA nor any other kind of molecule can, by itself, explain life.

Janus Among the Centaurs

The American architect Buckminster Fuller applied "synergy" (from Greek *synergos*, working together) to describe entities that behave as more than the sum of their parts. From a scientific standpoint, life, love, and behavior appear to be synergistic phenomena. When certain chemicals—in water and in oil—came together long ago, life was the result. Synergy also fits the emergence of protist cells from bacteria, and of animals from such cells.

The common view is that life evolves by random genetic change that is, moreover, detrimental more often than not. Chance mutations, blind and undirected, are touted to lead to evolutionary novelty. We (and a growing contingent of like-minded students of life) do not entirely agree. Great gaps in evolution have been leaped by symbiotic incorporation of previously refined components— components that have been honed in separate lineages. Evolution doesn't start anew each time a new life form appears. Preexisting modules (which turn out to be primarily bacteria), already generated by mutation and retained by natural selection, come together and interface. They form alliances, mergers, new organisms, whole new complexes that act and are acted on by natural selection.

But natural selection by itself cannot generate any evolutionary innovation, as Charles Darwin was well aware. Natural selection, rather, relentlessly preserves the former refinements and newly generated novelty by culling those less able to live or reproduce. Biotic potential—life's tendency to reproduce as much as possible—takes care of the rest. But first, novelty must arise from somewhere. In synergy two distinct forms come together to make a surprising new third one.

Cowboys, for example, settled the American West. Some native Americans perceived the human-horse invaders as centaurs—two-headed, multi-limbed beings. The novelist and philosopher Arthur Koestler (1905-1983) has called the coexistence of smaller beings in larger wholes "holarchy."[3] Most people, by contrast, think that life on Earth is hierarchical, a great chain of being with humans on top. Koestler's coinage is free of implications of "higher" or that one of the constituents in the holarchy is somehow controlling the others. The constituents, too, were given a new name by Koestler. Not merely parts, they are "holons"—wholes that also function as parts.

In his metaphysical as well as terminological rethinking, Koestler invoked the double-faced Janus, who in Roman mythology was the god of portals and the patron of beginnings and endings. In our view, just as Janus simultaneously looks backward and forward, so humans are not at the height of creation but point dually to the smaller realm of cells and the larger domain of biosphere. Life on Earth is not a created hierarchy but an emergent holarchy arisen from the self-induced synergy of combination, interfacing, and recombination.

Blue Jewel

The best part of a journey can be returning. By sending monkeys and cats into orbit, people to the moon, and robots to Venus and Mars, humankind has developed a new respect for, and a deeper understanding of life on Earth.

In 1961 the Soviet Union's *Vostok I* carried the first human into orbit around Earth. Since then, gazing "down" at this turquoise orb—venturing out on a spacewalk as if about to jump from the world's highest diving board—cosmonauts and astronauts have groped for words that do justice to their experience. Eugene A. Cernan, an astronaut of both the Gemini and Apollo lunar missions, and the last person to walk on the moon, describes the view:

> When you are in Earth orbit looking down you see lakes, rivers, peninsulas....You quickly fly over changes in topography, like the snow-covered mountains or deserts or tropical belts—all very visible. You pass through a sunrise and sunset every ninety minutes. When you leave Earth orbit...you can see from pole to pole and ocean to ocean without even turning your head....You literally see North and South America go around the corner as Earth turns on an axis you can't see and then miraculously Australia, then Asia, then all of America comes to replace them.... You begin to see how little we understand of time....You ask yourself, where am I in space and time? You watch the sun set over America and rise again over Australia. You look back "home"...and don't see the barriers of color, religion, and politics that divide up this world.[4]

Imagine yourself in orbit. Circling the planet every ninety minutes, the earthly experience of time and space undergo a mutual metamorphosis. Gravity lessens; up and down become relative. Day follows night in a patchwork blend. The sun cuts through the thin ribbon that is the atmosphere, flooding the cabin of the spacecraft from red to green to purple, through all the colors of the rainbow. You are plunged into black. Earth becomes the place where there are no stars. If Earth can be seen at all it is as a flicker of tiny lights—cities—on the surface of the sun-eclipsing globe. "Day" breaks again, revealing the cloud-flecked blue ocean. Jettisoned into a hyperperspective, the sky is now below. As if floating dreamily away from your own body, you watch the planet to which you are now tied by only the invisible umbilicus of gravity and telecommunication.

The act of viewing Earth from space echoes that of a baby glimpsing, and really seeing, itself in a mirror for the first time. [PLATE 5] The astronaut gazes upon the body of life as a whole. The French pyschoanalyst Jacques Lacan posits a stage in human development called "the mirror stage."[5] The infant, unable to control its limbs, looks into the mirror and perceives its whole body. Humanity's jubilant perception of the global environment represents the mirror stage of our entire species. For the first time we have caught a glimpse of our full, planetary form. We are coming to realize that we are part of a global holarchy that transcends our individual skins and even humanity as a whole.

Television images in 1969 revealed astronauts bounding over the lunar dust. The moon, once a synonym for the unattainable, was reached. A cratered wasteland, bone-dry, the moon was nevertheless still daunting in its lifelessness. As the cosmic perspective was broadcast, we homebodies were given a futuristic ride and were offered a new view of the world, a new worldview with the power to rally Earth's peoples around an icon more potent than any flag. Members of disparate religious and

A composite satellite photo showing visible evidence of night life on Earth: city lights. In orbit it becomes clear that night is shadow.

spiritual traditions could now join together as citizens of Earth. Individuals so affected, those who saw the potential, came to know that the whole former understanding of life was parochial, a result of where we lived. Even time was upset: night became shadow.

Tribal conflicts, national politics, and the colored geographic regions of maps are invisible from space. Science has, of course, revealed to us that this blue jewel orbits but a lackluster star in the outskirts of a spiral galaxy with myriad stars within a universe of myriad galaxies. All our history and civilization has transpired under the gaseous blanket of, really, a middling planet in one solar system. Voyaging in space, we saw Earth is home. But it is more than home: it is part of us. In contrast with the pale moon in the dead solar system of our galactic suburbs, this third planet from the sun, our Earth, is a blue-and-white flecked orb that looks alive.

Is There Life on Mars?

Unexpectedly, the search for life on Mars provided scientific confirmation of the "body" of life as a whole on Earth. The Viking mission, launched in 1975, sent two orbiters and two landers to Mars.

Although returning spectacular images of "Marscapes," the Viking landers performed a series of experiments that failed to find any evidence of Martian life. Channels carved by ancient rivers were seen, fueling hopes that evidence for past life may yet be found on the red planet.

One scientist, however, was able to search for life on Mars before the Viking mission was launched. In 1967 James E. Lovelock, English inventor of a device that measures chlorofluorocarbons implicated in the production of ozone holes, was consulted by the National Aeronautics and Space Administration (NASA) in its search for extraterrestrial life. NASA was interested in what Lovelock's invention, a gas-measuring instrument some thousand times more sensitive to certain atmospheric constituents than any previous device, might reveal about Mars. An atmospheric chemist, Lovelock suspected that, in principle, life on any planet could be detected by the chemical markers left in that planet's air. Because the constituents of Mars's atmosphere were already known by the spectroscopic signature of the planet's reflected light, Lovelock believed the data already sufficient to determine whether Mars was a living planet. His conclusion: Mars was devoid of life.

Indeed, he boasted with his own brand of quiet iconoclastic mischief that his prediction precluded any need to visit Mars at all and that he could save NASA a prodigious sum of money.

Lovelock had measured Earth's atmospheric gases with a chromatograph outfitted with his new supersensitive "electron capture device." He was startled: the chemistry of Earth's atmosphere, not at all like the atmospheres of Mars and Venus, is utterly improbable. He found that methane, a chief constituent of natural gas and present in the atmospheres of the four giant planets (Jupiter, Saturn, Uranus, and Neptune), freely coexisted in Earth's atmosphere with oxygen at concentrations 10^{35} times higher than expected.

Methane exists at only one to two parts per million in Earth's atmosphere, but even that minuscule proportion is far too high. Methane (one carbon atom surrounded by four hydrogen atoms) and oxygen gas (two oxygen atoms) react explosively with each other to generate heat, producing carbon dioxide and water. Oxygen, the second most abundant gas in the atmosphere, should thus react immediately with methane to make the latter undetectable. Perhaps in the next minute you will die of asphyxiation because all the oxygen atoms will gather in one corner of the room and your brain will be deprived of its absolute requirement for oxygen gas. Such a calamity is improbable to the point of absurdity. Yet the chemical mixture of methane and oxygen in the Earth's air is equally freakish. Indeed, not only methane but many other gases in our air should not be detectable, given standard rules of chemical mixing. Given their tendency to react with oxygen, some of our atmosphere's components—methane, ammonia, sulfur gases, methyl chloride, and methyl iodide—are far from chemical equilibrium. Carbon monoxide, nitrogen, and nitrous oxide are respectively ten, ten thousand million, and ten trillion times more abundant than chemistry alone can account for.

Biology, however, offers an answer. Lovelock realized that methane-producing bacteria release this gas in globally significant amounts. Cows, for instance, contribute methane by belching. Belched methane does react with oxygen but, before it disappears, more is produced. The methane is made from grass by protists and bacteria in the cow's rumen, a special stomach.

Life has made our atmosphere chemically reactive and orderly, while exporting heat and disorder to space. Lovelock maintained that the atmosphere is as highly ordered as a painted tortoise's shell or a sand castle on a deserted beach. And life's inveterate ordering has left its traces on other planets. On 20 July 1976 a lander was left on Mars by the 3.6 metric ton *Viking I* spacecraft. [PLATE 6] Although not what scientists were looking for, this machine, sitting 571 million kilometers away at Chryse Planitia on red sand, is the best, and so far the only evidence of life on Mars: solar system exploring, technological human life.

Life as Verb

Lovelock's analyses have pushed biologists to realize that life is not confined to the things now called organisms. Self-transforming, holarchic life "breaks out" into new forms that incorporate formerly self-sufficient individuals as integral parts of greater identities. The largest of these levels is the planetary layer, the biosphere itself. Each level reveals a different kind of "organic being." This is the term that Darwin used throughout his opus, *The Origin of Species*. ("Organism," like "scientist" and "biology," had not yet been coined.) "Organic being" merits resurrection as it affords the recognition that a "cell" and the "biosphere" are no less alive than an "organism."

Life—both locally, as animal, plant, and microbe bodies, and globally, as the biosphere—is a most intricate material phenomenon. Life shows the usual chemical and physical properties of matter, but with a twist. Beach sand is usually silicon dioxide. So are landed September 3, 1976. The blue-star field and red stripes of the USA's flag help corroborate the chromatic correctness of the salmon-colored sky caused by dust particles suspended in the Martian atmosphere.

the innards of a mainframe computer—but a computer isn't a pile of sand. Life is distinguished not by its chemical constituents but by the behavior of its chemicals. The question "What is life?" is thus a linguistic trap. To answer according to the rules of grammar, we must supply a noun, a thing. But life on Earth is more like a verb. It repairs, maintains, recreates, and outdoes itself.

This surge of activity, which not only applies to cells and animals but to Earth's entire atmosphere, is intimately connected to two of science's most famous laws—the laws of thermodynamics. The first law says that throughout any transformation the total energy of any system and its environment is neither lost nor gained. Energy—whether as light, movement, radiation, heat, radioactivity, chemical or other—is conserved.

But not all forms of energy are equal; not all have the same effect. Heat is the kind of energy to which other forms tend to convert, and heat tends to disorganize matter. The second law of thermodynamics says that physical systems tend to lose heat to their surroundings.

The second law was conceived during the Industrial Revolution, when the steam engine represented the state-of-the-art in engineering. French physicist Nicolas Carnot (1796–1832), aiming to improve the efficiency of the steam engine (whose governor mechanism was invented by James Watt), came to realize that heat was associated with the movement of minute particles. And from that, he envisioned the principle that is now known as the second law: In any moving or energy-using system entropy increases.

In systems undergoing change, such as steam engines or electric motors, a certain amount of the total energy available is already in (and more is converted into) a form that is unavailable for useful work. Although the amount of energy in the system and its environment stays the same (i.e., the first law of thermodynamics, of conservation of energy,

holds), the amount of energy available to do work decreases. In computer science entropy is measured as the uncertainty in the information content of a message. The second law unequivocally claims that in changing systems entropy increases, implying that heat, noise, uncertainty, and other such forms of energy not useful for work, increase. As local systems lose heat, the universe as a whole is gaining it. Although not so popular now, in the past physicists and chemists have made the prediction that the universe will whimper out in a "heat death" as a consequence of the tendency for entropy to increase. More recently, they have even invented the word "negentropy" for life, which, in its tendency to increase information and certainty, seems to contradict the second law. It doesn't; the second law holds as long as one regards the system (life) in its environment.

In steam engines, coal was burned and carbon joined with oxygen, a reaction that, generating heat, made machine parts move. The left-over heat that was generated was unusable. The heat in a cabin on a snow-covered mountain seeks with seeming purpose any available crack or opening to mix with the cold air outside. Heat naturally dissipates. This dissipative behavior of heat illustrates the second law: the universe tends toward an increase in entropy, toward even temperatures everywhere, as all the energy transforms into useless heat spread so evenly that it can do no work. Heat dissipation, we are usually told, results from random particle motion. But there are other interpretations.

Some scientists have begun to interpret the second law's predilection for heat-energy as the basis for apparent purposeful action. Ilya Prigogine, a Belgian Nobel laureate, helped pioneer the consideration of life within a larger class of "dissipative structures," which also includes decidedly nonliving centers of activity like whirlpools, tornadoes, and flames.[6] A rather awkward term because it focuses on what the structures—actually, systems, not structures—throw away rather than what they retain and build, a dissipative system maintains itself (and may even grow) by importing "useful" forms of energy and exporting, or dissipating, less useful forms—notably, heat. This thermodynamic view of life actually goes back to Schrödinger, who also likened living beings to flames, "streams of order" that maintain their forms.

American scientist Rod Swenson has argued that the seeming purpose displayed in heat's tendency to dissipate with time is intimately related to the behavior of life forms striving to perpetuate themselves. In Swenson's view, this entropic universe is pocked by local regions of intense ordering (even life) because it is through ordered, dissipative systems that the rate of entropy production in the universe is maximized. The more life in the universe, the faster that various forms of energy are degraded into heat.[7]

Swenson's view shows how life's seeming purpose—its seeking behavior, its directedness, which philosophers call teleology—is related to the behavior of heat. Scientists do not as a rule endorse teleology. They consider it unscientific, a holdover from the primitive days of animism. Teleology is nevertheless embedded in language, and it cannot and need not be eliminated from the sciences. The prepositions "to" and "for," which build teleology—that is, purposefulness—into language, speak of a future-directedness that seems present, to some degree, in all living beings. One should not assume that only humans are future-oriented. Our own frenetic attempts (and those of the rest of life) to survive and prosper may be a special, four-billion-year old way the universe has organized itself "to" obey the second law of thermodynamics.

Self-Maintenance

Islands of order in an ocean of chaos, organisms are far superior to human-built machines. Unlike James Watts's steam engine, for example, the body concentrates order. It continuously self-repairs. Every five days you get a new stomach lining. You get a new liver every two months. Your skin replaces itself every six weeks. Every year, ninety-eight percent of the atoms of your body are replaced. This nonstop chemical replacement, metabolism, is a sure sign of life. This "machine" demands continual input of chemical energy and materials (food).

Chilean biologists Humberto Maturana and Francisco Varela see in metabolism the essence of something quite fundamental to life. They call it "autopoiesis." Coming from Greek roots meaning self (*auto*) and making (*poiein*, as in "poetry"), autopoiesis refers to life's continuous production of itself.[8] Without autopoietic behavior, organic beings do not self-maintain—they are not alive.

An autopoietic entity metabolizes continuously; it perpetuates itself through chemical activity, the movement of molecules. Autopoiesis entails energy expenditure and the making of messes. Autopoiesis, indeed, is detectable by that incessant life chemistry and energy flow which is metabolism. Only cells, organisms made of cells, and biospheres made of organisms are autopoietic and can metabolize.

DNA is an unquestionably important molecule for life on Earth, but the molecule itself is not alive. DNA molecules replicate but they don't metabolize and they are not autopoietic. Replication is not nearly as fundamental a characteristic of life as is autopoiesis. Consider: the mule, offspring of a donkey and a horse, cannot "replicate." It is sterile, but it metabolizes with as much vigor as either of its parents; autopoietic, it is alive. Closer to home, humans who no longer, never could, or simply choose not to reproduce can not be relegated, by the strained tidiness of biological definition, to the realm of the nonliving. They too are alive.

In our view, viruses are not. They are not autopoietic. Too small to self-maintain, they do not metabolize. Viruses do nothing until they enter an autopoietic entity: a bacterial cell, the cell of an animal, or of another live organism. Biological viruses reproduce within their hosts in the same way that digital viruses reproduce within computers. Without an autopoietic organic being, a biological virus is a mere mixture of chemicals; without a computer, a digital virus is a mere program.

Smaller than cells, viruses lack sufficient genes and proteins to maintain themselves. The smallest cells, those of the tiniest bacteria (about one ten-millionth of a meter in diameter) are the minimal autopoietic units known today. Like language, naked DNA molecules, or computer programs, viruses mutate and evolve; but, by themselves, they are at best chemical zombies. The cell is the smallest unit of life.

When a DNA molecule produces another DNA molecule exactly like itself, we speak of replication. When living matter, as a cell or as a body made of cells, grows another similar being (with differences attributable to mutation, genetic recombination, symbiotic acquisition, developmental variation, or other factors), we speak of reproduction. [PLATE 7] When living matter continues to reproduce altered forms that, in turn, make altered offspring, we speak of evolution: change in populations of life forms over time. As Darwin and his legacy stress, more reproducing cells and bodies are produced by budding, cell division, hatching, birth, spore forma-tion, and the like than can ever survive. Those that cope long enough to reproduce are "naturally selected." More bluntly, survivors are not so much selected for their success as those failing to repro-duce before dying are selected against.

Identity and self-maintenance require metabolism. Metabolic chemistry (often called physiology) precedes reproduction and evolution. For a population to evolve, its members must reproduce. Yet before any organic being can reproduce, it must first self-maintain. Within the lifetime of a cell, each of five thousand or so different proteins will completely interchange with

7
Pachnoda. Phylum: Arthropoda. Kingdom: Animalia. Close-up of the intestine of the beetle larva showing a tree-like organ that houses methanogenic bacteria. Living in the intestine for thousands of generations, the methangens have not only found a home but produced this symbiotic "beetle" organ.

the surroundings thousands of times. Bacterial cells produce DNA and RNA (nucleic acids), enzyme proteins, fats, carbohydrates, and other complex carbon chemicals. Protist, fungi, animal, and plant bodies all produce these and other substances as well. But most importantly, and amazingly, any living body produces itself.

This energetic maintenance of unity while components are continuously or intermittently rearranged, destroyed and rebuilt, broken and repaired is metabolism, and it requires energy. In accordance with the second law, autopoietic self-maintenance preserves or increases internal order only by adding to the "disorder" of the external world, as wastes are excreted and heat is vented. All living beings must metabolize and therefore all must create local disorder: useless heat, noise, and uncertainty. This is autopoietic behavior, reflecting the autopoietic imperative required for any organic being that lives, that continues to function.

The autopoietic view of life differs from standard teachings in biology. Most writers of biology texts imply that an organism exists apart from its environment, and that the environment is mostly a static, unliving backdrop. Organic beings and environment, however, interweave. Soil, for example, is not unalive. It is a mixture of broken rock, pollen, fungal filaments, ciliate cysts, bacterial spores, nematodes, and other microscopic animals and their parts. "Nature," Aristotle observed, "proceeds little by little from things lifeless to animal life in such a way that it is impossible to determine the exact line of demarcation."[9] Independence is a political, not a scientific, term.

Since life's origin, all living beings, directly or circuitously, have been connected, as their bodies and populations have grown. Interactions occur, as organisms connect via water and air. Darwin, in his *Origin of Species*, likened the complexity of these interactions to "an entangled bank"—too complex for us humans even to begin to sort out: "Throw up a handful of feathers, and all fall to the ground according to definite laws; but how simple is the problem where each shall fall compared to that of the action and reaction of the innumerable plants and animals." Yet it is the sum of these uncountable interactions that yields the highest level of life: the blue biosphere, in all the holarchic coherence and mysterious grandeur of its budding in and from the black cosmos.

The Autopoietic Planet

The biosphere as a whole is autopoietic in the sense that it maintains itself. One of its vital "organs," the atmosphere, is clearly tended and nurtured. Earth's atmosphere, approximately one-fifth oxygen, differs radically from those of Mars and Venus. The atmospheres of these planetary neighbors are nine parts in ten carbon dioxide; in Earth's atmosphere, carbon dioxide accounts for only three parts in ten thousand. If Earth's biosphere were not made of carbon dioxide-consuming beings (plants, algae, photosynthetic, and methane-producing bacteria, among myriad other life forms), our atmosphere would long ago have reached a carbon dioxide-rich chemical balance. And virtually every molecule capable of reacting with another molecule would already have reacted. Instead, the combined activities of autopoietic surface life have led to an atmosphere in which oxygen has been maintained at levels of about twenty percent for at least 700 million years. [CHART A]

Other evidence for life on a planetary scale comes from astronomy. According to standard astrophysical models of the evolution of stars, the sun used to be cooler than it is now. The sun's luminosity has increased by thirty percent or more since life began on Earth. Living things can grow and reproduce only in a limited temperature range within which water is liquid. Fossils of life more than three thousand million years old confirm that ancient temperatures were not all that dissimilar

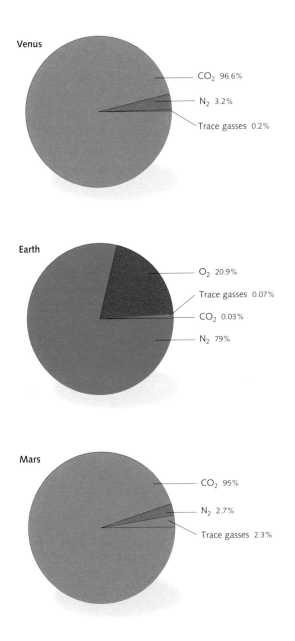

Venus

CO$_2$ 96.6%

N$_2$ 3.2%

Trace gasses 0.2%

Earth

O$_2$ 20.9%

Trace gasses 0.07%

CO$_2$ 0.03%

N$_2$ 79%

Mars

CO$_2$ 95%

N$_2$ 2.7%

Trace gasses 2.3%

Chart A

Atmospheric comparison of Earth and its two planetary neighbors. Note the comparatively high concentration of the explosive gas oxygen and the very low concentration of carbon dioxide on Earth.

This atmospheric anomaly results from the incessant activity of gas-exchanging organisms. The minute physiology of the cell over geological time becomes magnified into the global physiology of the biosphere.

from those prevailing today; other geological evidence suggests that liquid water was widespread on Earth at least four thousand million years ago. The increase in the luminosity of the sun should have dramatically increased the surface temperature of Earth since those early times. Because no dramatic increase has occurred—indeed, the trend may have been a cooling—it appears that the temperature of the entire biosphere has been self-maintained. By responding, life seems to have succeeded in cooling the planetary surface to counter, or more than counter, the overheating sun. Mainly by removing from the atmosphere greenhouse gases (such as methane and carbon dioxide) which trap heat, but also by changing its surface color and form (by retaining water and growing slime), life responded to prolong its own survival.

Oceanography provides still another glimpse of the body of life as a whole. Chemical calculations suggest that salts should accumulate in the oceans to concentrations perilous to nonbacterial forms of life. Salts, such as sodium chloride and magnesium sulfate, are continuously eroded from the continents and carried into the oceans by rivers. World oceans have, however, remained hospitable to salt-sensitive organisms for at least two thousand million years. Seafaring microorganisms may therefore be sensing and stabilizing ocean acidity and salinity levels on a global scale. How life removes salt from marine waters is obscure. Perhaps salt concentrations too high for most life are lowered, in part, by the vigorous pumping of sodium, calcium, and chloride out of cells and, in part, by formation of evaporite flats. These encrusted fields are rich in sea salt and salt-loving microbes. They often form behind lagoonal barriers made by animals such as corals or when shifting sands are trapped by the mucus and slime formed by microbial communities. Continuous desalination, if it exists, may be part of a global physiology.

Some evolutionary biologists have suggested that

Earth life in its totality cannot constitute a living body, cannot be a living being, because such a body could only have evolved in competition with other bodies of the same sort—presumably, other biospheres. But, in our view, autopoiesis of the planet is the aggregate, emergent property of the many gas-trading, gene-exchanging, growing, and evolving organisms in it. As human body regulation of temperature and blood chemistry emerges from relations among the body's component cells, so planetary regulation evolved from eons of interactions among Earth's living inhabitants.

Using the energy of sunlight, only green plants, algae, and certain green- and purple-colored bacteria can convert compounds from surrounding water and air into the living stuff of their bodies. This sun-energized process, photosynthesis, is the nutritional basis for all the rest of life. Animals, fungi, and most bacteria feed on the purple and green producers. Photosynthesis evolved in microbes soon after the origin of life. At every level, from microbe to planet, organic beings use air and water or other organic beings to build their reproducing selves. Local ecology becomes global ecology. As a corollary, and in spite of English grammar, life does not exist *on* Earth's surface so much as it *is* Earth's surface.

Life extends over the planet as a contiguous, but mobile, cover and takes the shape of the underlying Earth. Life, moreover, enlivens the planet; Earth, in a very real sense, is alive. This is no vague philosophical claim but rather a physiological truth of our lives. Organisms are less self-enclosed, autonomous individuals than communities of bodies exchanging matter, energy, and information with others. Each breath connects us to the rest of the biosphere, which also "breathes," albeit at a slower pace. The biosphere's breath is marked daily by increasing carbon dioxide concentrations on the dark side of the globe and decreasing concentrations on the

lighted side. Annual breathing is marked by the passage of the seasons; photosynthetic activity kicks up in the northern hemisphere just as it is winding down in the Southern.

Taken at its greatest physiological extent, life *is* the planetary surface. Earth is no more a planet-sized chunk of rock inhabited with life than your body is a skeleton infested with cells. [CHART B]

Seasonal Fluctuations of Carbon Dioxide in the Northern Hemisphere
(Measured in parts per million.)

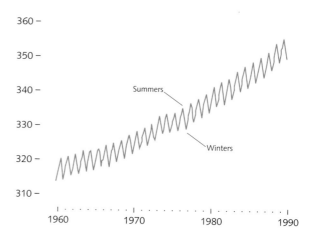

Chart B
The peaks of the zigzags represent increase in atmospheric carbon dioxide during summers in the northern hemisphere; the overall upward trend indicates rising levels of CO_2 due to industrial activity. This seasonal and annual fluctuation of car- bon dioxide in the Earth's atmosphere attests to "breathing" on a global scale. The total carbon dioxide increase may, by the greenhouse effect, raise planetary tempera- tures to levels inhospitable for human beings—a geophysiological "fever."

The Stuff of Life

When German chemist Friedrich Wöhler (1800-1882) first, accidentally, produced crystals of urea by heating ammonium cyanate, he could not accept that he had made from scratch a compound so clearly associated with living beings. Urea, after all, is the carbon-nitrogenous waste produced in animal urine. And in Wöhler's day, organic beings were believed to consist of a strange and wonderful "organic matter" that was present in life—and nowhere else. Since then, dozens of carbon-rich compounds, such as formic acid, ethylene, and hydrogen cyanide, have been found not just in life but in interstellar space. The equivalent of an estimated 10 quintillion (10,000,000,000,000,000,000) fifths of whiskey, in the form of the nine-atom molecule CH_3CH_2OH (ethyl alcohol), exists in one interstellar cloud in the constellation of Orion alone.

Though adulterated with other compounds, we, like all living matter, are mostly water—that is, hydrogen and oxygen. Hydrogen forms, by mass, seventy-five percent of the atoms in the cosmos. It is the same element which, under intense gravitational pressure, becomes helium in the nuclear fusion reaction that makes our sun shine. Far older and bigger stars went out with a bang, as supernovas, and thereby created carbon, oxygen, nitrogen, and the other heavier elements. Life is made from such star stuff. In the universe life may be rare or even unique. But the stuff of which it is made is readily available.

More and more inert matter, over time, has literally come to life. Minerals of the sea are now incorporated into living creatures for protection or support in the form of integument, shell, bone. Our own skeletons are built from calcium phosphate, a sea salt that was initially a nuisance or a hazard for our remote ancestors, marine protist cells which eventually found ways to cleanse their tissues by putting such minerals to use. The kinds as well as the mass of chemical elements in living bodies have increased through evolutionary time. Whereas structural compounds made of hydrogen, oxygen, sulfur, phosphorus, nitrogen, and carbon are required by all cells and have been essential to life since its inception, those made of silicon and calcium are relative newcomers.

Heinz Lowenstam (1913-1993), a Silesian-born geologist and refugee from Nazi Germany, cataloged the minerals produced in the hard parts of animals. In Lowenstam's youth, the only hard substances thought to be produced by living tissues were the calcium phosphate of our own bones and teeth, the calcium carbonate of mollusk shells, and the silicon dioxide of unusual structures such as the spicules of sponge. [PLATES 8 AND 9, *following page*] Lowenstam and his colleagues went on to discover many other minerals produced by life. The list of hard substances made by live cells, including unexpectedly beautiful crystals, now surpasses fifty. [CHART C, *following page*]

Life had been reusing hard materials and shaping solid wastes long before the appearance of technological humans. Bacteria came together to form protoctists that in turn could mine and use calcium, silica, and iron from the world's seas. Protoctists evolved into animals with shells and bones. Animals, individually or in concert, engineered inert materials into tunnels, nests, hives, dams, and the like. Even some plants incorporate minerals. The silica-laced bodies of "scouring rushes," for example, may serve as good pot scrubbers for campers, but they have probably evolved to deter herbivores. The calcium oxalate crystals of *Dieffenbachia* are hurtled from the leaf cells toward unwary, hungry victims.

The propensity to "engineer" environments is ancient. Today people make over the global environment. Clothed and bespectacled inside an automobile, connected by phone wires and radio

8

Oxalic acid crystal taken from a sea squirt renal sac, an organ thought to be a ductless kidney. *Nephromyces*, a protist probably associated with symbiotic bacteria, apparently forms the crystals from the animal's uric acid and calcium oxalate. Over 50 such minerals are now known to be produced in living cells.

9

Sea squirts. Phylum: Chordata. Kingdom: Animalia. Some of these sessile sea animals (also called didemnids or sea lemons) produce calcium oxalate crystals such as that shown in plate 8.

Chart C

Contrary to popular belief minerals and animals do not belong to separate kingdoms. Many minerals are produced in and by life, sometimes in crystalline form. One of the most common minerals, calcium carbonate, is formed by living marine animals as shells. Another compound, calcium phosphate, is precipitated by cells of our bones. As this table shows all five kingdoms of organisms have members which produce minerals. This list represents only a sample of the over fifty minerals now known to be produced by living cells.

Mineral	Bacteria	Protoctista	Fungi	Animals	Plants
CALCIUM					
Calcium carbonate ($CaCO_3$; aragonite, calcite, vaterite)	SHEATH AND OTHER EXTRACELLULAR PRECIPITATE	AMOEBA AND FORAMINIFERAN SHELLS	EXTRACELLULAR PRECIPITATES MUSHROOMS	CORALS MOLLUSK SHELLS ECHINODERM SKELETONS CALCAREOUS SPONGES SOME KIDNEY STONES	EXTRACELLULAR PRECIPTATES
Calcium phosphate ($CaPO_4$)			EXTRACELLULAR PRECIPITATES MUSHROOMS	BRACHIOPOD "LAMP SHELLS" VERTEBRATE TEETH AND BONES SOME KIDNEY STONES	
Calcium oxalate (CaC_2O_4)				MOST KIDNEY STONES	*DIEFFENBACHIA*, A FLOWERING PLANT
SILICON					
Silica (SiO_2)	PRECIPITATES	DIATOM AND RADIOLARIAN SHELLS MASTIGOTE ALGAE SCALES		GLASS-SPONGE SPICULES	GRASS PHYTOLITHS HORSE-TAIL STEMS
IRON					
Magnetite (Fe_3O_4)	MAGNETOSOMES			ARTHROPODS MOLLUSKS VERTEBRATES	
Greigite (Fe_3S_4)	MAGNETOSOMES				
Siderite ($FeCO_3$)	EXTRACELLULAR PRECIPITATES				
Vivianite ($Fe_3[PO_4]_2 \cdot 8H_2O$)	EXTRACELLULAR PRECIPITATES				
Goethite ($xxFeO \cdot OH$)	EXTRACELLULAR PRECIPITATES		EXTRACELLULAR PRECIPITATES	CHITONS	
Lepidocrocite ($xxFeO \cdot OH$)	EXTRACELLULAR PRECIPITATES		EXTRACELLULAR PRECIPITATES MUSHROOMS	CHITONS	
Ferrihydrite ($5Fe_2O_3 \cdot 9H_2O$)				MOLLUSKS	FLOWERING PLANTS
MANGANESE					
Manganese dioxide (MnO_2)		INTRACELLULAR OR EXTRACELLULAR PRECIPITATES AROUND SPORES			
BARIUM					
Barium sulfate ($BaSO_4$)		ALGAL-PLASTID GRAVITY SENSORS MARINE PROTIST SKELETONS		SENSE ORGANS: STATOLITHS (OTOLITHS)	
STRONTIUM					
Strontium sulfate ($SrSO_4$)		MARINE PROTIST SHELLS		MOLLUSK SHELLS	

waves to modems, cellular phones, and bank machines, supplied with electricity, plumbing, and other utilities, we are transforming ourselves from individuals into specialized parts of a global more-than-human being. This metahuman being is inextricably bound to the much older biosphere, from which it arose. Metals and plastics represent the newest realm of matter "coming to life."

Mind in Nature

The biological self incorporates not only food, water, and air—its physical requirements—but facts, experiences, and sense impressions, which may become memories. All living beings, not just animals but plants and microorganisms, perceive. To survive, an organic being must perceive—it must seek, or at least recognize, food and avoid environmental danger.

A living being need not be conscious to perceive. But consider: most of our own daily activities (breathing, digesting, even turning a page or driving a car) are performed largely or even wholly unconsciously. From the viewpoint of the evolutionary biologist, it is reasonable to assume that the sensitive, embodied actions of plants and bacteria are part of the same continuum of perception and action that culminates in our own most revered mental attributes. "Mind" may be the result of interacting cells.

Mind is fully an evolutionary phenomenon. Hundreds of millions of years before organic beings verbalized life, they recognized it. Discerning what could kill them, what they could eat, and what they could mate, roughly in that order, were crucial to animal survival. One U.S. Supreme Court justice avowed that while he might not be able to define obscenity he surely could recognize it when he saw it. We all have a similar ability with life. Life has been recognizing itself long before any biology textbooks were written.

Survival-based psychological tendencies infiltrate the pristine realm of science. Pattern recognition was such a useful trait for our ancestors that, even if occasionally wrong, the *Aha!* feeling of discovery would have been reinforced. Aesthetic judgments of elegance and beauty, often cited in the preference for certain equations over others in physics, show that scientific correctness can also be intuitive. What we know, what we are capable of knowing and seeing, has been shaped by our evolution as surviving creatures. Even foolish and outlandish notions would have been retained and reinforced if they in any way aided our ancestors' survival.

Neuroscientists have traced subjective feelings of pleasure to endorphins and enkephalins, two groups of neuropeptides produced by the brain. The pleasure associated with seeing beauty, including scientific "truth," may have come about during the course of evolution, just as love and biophilia—the pleasure we take in the company of other living creatures—provoke us to seek out mates and the natural environments that have been most conducive to our survival. If we did not fear death, we might be too quick to kill ourselves when troubled or inconvenienced and thus perish as a species. Belief in life's importance may not be a reflection of reality, then, but an evolutionarily reinforced fantasy that prejudices believers to do what is necessary, bear whatever burdens, to survive.

We all inherit a shared perspective bequeathed by our ancestors. The physicists' hope of solving an essential set of equations for all time and the cosmos may be but the gleam of a receding mirage. In the end, as Charles Peirce and William James recognized, there may be no better measure of "truth" than that which works—that which helps us survive.

Mind and body, perceiving and living, are equally self-referring, self-reflexive processes already present in the earliest bacteria. Mind, as well as body, stems from autopoiesis. And in sufficiently expressive humans the process of autopoiesis underlying living organization makes itself manifest even outside the body. Abstract expressionist painter Willem de Kooning wrote:

> If you write down a sentence and you don't like it, but that's what you wanted to say, you say it again in another way. Once you start doing it and you find how difficult it is, you get interested. You have it, then you lose it again, and then you get it again. You have to change to stay the same.[10]

Changing to stay the same is the essence of autopoiesis. It applies to the biosphere as well as the cell. Applied to species, it leads to evolution.

So, what is life?

It is a material process, sifting and surfing over matter like a strange, slow wave. It is a controlled, artistic chaos, a set of chemical reactions so staggeringly complex that more than eighty million years ago it produced the mammalian brain that now, in human form, composes love letters and uses silicon computers to calculate the temperature of matter at the origin of the universe. Life, moreover, appears to be on the verge of perceiving for the first time its strange but true place in an inexorably evolving cosmos.

Life, a local phenomenon of Earth's surface, can in fact be understood only in its cosmic milieu. It formed itself out of star stuff, shortly after Earth 4,600 million years ago congealed from a remnant of a supernova explosion. Life may end in a mere hundred million years when, embattled by dwindling atmospheric resources and increased heat from the sun, systems of global temperature regulation finally fail.[11] Or life, enclosed in ecological systems, may escape and watch from safe harbor as the sun, exhausting its hydrogen, explodes into a red giant, boiling off Earth's oceans, five thousand million years from now. ■

2

Lost Souls

Ay, but to die, and go we know not where;
To lie in cold obstruction and to rot.

—**William Shakespeare,** *Measure for Measure*

Love that endures for a breath:
Night, the shadow of light,
And life, the shadow of death.

—**Algernon Swinburne,** *Atalanta in Calydon*

Death: The Great Perplexer

The scientific mystery of life in a near-lifeless, mechanical universe mirrors the enigma of death in a fully living, animistic one. Our ancestors inhabited a world where warm, moving bodies would regularly stop, grow cold, and decay. As puzzling as life is for us, so was death for them. But we moderns still feel the influence of ancient solutions to the death puzzle.

Until the seventeenth century the sun and moon did not move according to Newtonian principles; these celestial bodies often were animated by spirits within them. The whistling of the wind, the changing phases of the moon, the twinkling, turning stars—these eternal, celestial bodies moved as they

Burial of two *Homo sapiens*, possibly a young woman and infant, in the Qafzeh cave in present-day Israel. Such burial sites, often replete with jewelry, medicines, and other possessions of no benefit to decaying corpses, attest to an ancient belief in a spirit world. These bodily remains are 100,000 years old.

willed to move, as we move by will. But what happened to the will of the warrior, whose heart had beat so hotly a moment ago, and is now a cold corpse? Does life slip away at spear point in a pool of blood? As the corpse becomes rigid, does the enlivening spirit dart into the grass? Vanish into thin air?

Originally death, not life, was the great perplexer.

In a living universe what is death? Where do "we" go when we die? A gold coin disappears in one hand as the magician produces another just like it in the opposite. The spectator concludes the gold migrated from hand to hand, just as the logical mind concludes that the soul sneaks out of the body after death.

Any nearby being could have taken the missing soul. An infant, a goat, a snake—a raven at the scene of the crime—could have snatched the essence whose lack rendered a body lifeless.

An apparent attention to the mystery of death marks the earliest human remains. Sixty thousand years ago, at Shanidar cave in Iraq, a Neanderthal man was buried on a mat of woven pine boughs and with flowers related to grape hyacinth, bachelor's button, hollyhock, and groundsel.[1] Such grave sites, filled with flower parts, pollen, amulets, beads, headbands made of fox teeth, weapons, tools, and food, attest to funerary rites seemingly designed to provide the soul with rest—and with the goods it would require in an afterlife. [PLATE 10]

The Breath of Life

No great leap of faith is needed to see how early mysteries of the corpse led to religious notions of spirit. For the Iroquois of North America, the spirit was an exceedingly refined image, possessing a tiny body replete with head, teeth, and limbs. The Karo Battak of Sumatra envisioned a "tendi," a copy of the owner or other self, which flees at death. The peoples of Papua and Malay posited a brown, corn kernel-sized *semangat* or *semungi*, which, if it departed temporarily, induced malady and, if permanently, death. Even the inventor of the microscope, Antoni van Leeuwenhoek (1632-1723), thought he saw the homunculus—a tiny, human-shaped seed—when he examined sperm.

Some cultures have regarded blood, others flesh, (the Australians, the kidney fat) as the seat of life. New Zealand Maoris held to the notion that the menstrual blood is the source of life. Shadows, flames, trees, columns, dolls, pools, children, and Polaroid photographs have all been proposed to detain or permanently entrap souls.

Of the candidates for life's essence, breath is strongest. The ancient Chinese used durable, airtight coffins of cypress and pine, tightly binding and stuffing the mouths of their dead with jade, gold, silver, pearls, and cowry shells to restrain the spirit. The word spirit itself comes from *spiritus*, Latin for breath. Birth is announced by crying—and breathing. So long as there is life, there is breath.

Breath is invisible. Like wind, it moves things. Moreover, we speak with breath. Shamans and priests from many cultures concluded that air as spirit, perhaps the holy spirit of some unseen breathing being, was the intangible link between life and death. Witness the etymological kinship between inspiration, expiration, respiration, and spirit. Then too, expiration is itself a synonym for death. In many native American languages Great Spirit and Great Wind share the same word and meaning. The Aztec word *ehecatl* means wind, air, life, soul, shadow. A common term in the *Old Testament*, *nephesh* means living spirit or breath-soul, and to die is to "breathe out the *nephesh*." The Chinese *ch'i*, important both to martial arts and medicine, is the life force, the cosmic spirit that pervades and enlivens all things and that is synonymous with primordial energy. Although for the ancient Greeks *psyche* meant "breath-soul" (as distinct from "blood-soul," which was the seat of consciousness), by Aristotle's time psyche had come to mean the life principle. *Pneuma*, another Greek term for spirit or soul, familiar to us in "pneumonia" or "pneumatic," derives from the word *pnein*, which means to breathe. In *De Anima* ("On the Soul") Aristotle claimed the soul, the purpose for which the living body exists, is the source of motion ("animation").

The magico-religious notion of a holy spirit breathing souls into bodies worked its way into science. Before the eighteenth century live beings were not thought to "reproduce"; they were "generated." The monsters of bestiaries were the combinatorial results of souls and natural and divine intervention. Aristotle thought that by way of semen men contribute souls whereas women provide the nurturing matter in the generation of the child. "Parents are merely the seat of the forces uniting matter and form," wrote Jean François Fernel (1497-1558), who coined the terms physiology and pathology and who was physician to King Henry II of France. "Above them, stands a more powerful Workman. It is He who determines the form by breathing the breath of life."[2]

The observation that some things, such as rocks, lacked animating souls implied the obverse, that self-moving spirits floated bodilessly thru the ether. Combined with a longing for immortality, this inference—that souls exist on their own—offered hope of cheating death. The notion of disembodied spirit is at the root of ancestor worship and beliefs in ghosts, angels, and reincarnation. For Plato the heavens were inhabited by ensouled planets and stars, the world a divine repetition, within time, of a perfect realm beyond time: a universe of pure mind. In *The Timaeus* he wrote, "The world has received animals, mortal and immortal, and is fulfilled with them, and has become a visible animal containing the visible—the sensible God who is the image of the intellectual, the greatest, best, fairest, most perfect—the one only-begotten heaven." Aristotle, who studied living beings well enough to correctly describe the use of tentacles in mating octopi, modified Plato's ideas. He emphasized the mundane purpose of living beings as the great purpose generated by "First Cause" or "Unmoved Mover." Christianity, influenced by Greek philosophy through the Church fathers, incorporated the Hebrew notion of a single God. Christian doctrine dispensed with nature spirits and ancillary gods save those, such as saints and angels, who mediate between man, his soul, and God. The souls and spirits once perceived to be rampant throughout nature became vanishingly scarce.

In the Middle Ages (c. 500-1500 C.E.) a European religious sect known as the Gnostics decided that the true self was a divine spark trapped in a prison of fleshly matter. The Gnostics pictured Earth as surrounded by seven transparent, crystalline spheres—the clear heavens—each of which held a celestial body: Moon, Mercury, Venus, Mars, Jupiter, Saturn, and Sun. The spheres holding the planets were living spiritual powers, three-dimensional glass ceilings, so to speak, presided over by Archons, cosmic bouncers whose business it was to prevent souls from rising back to heaven. Gnosticism made sense in a medieval Europe. The Black Plague—half-dead bodies moaning and rotting in the streets, warmer ones flagellating one another and predicting the end of the world—called into question the behavior of a Creator no less than the Holocaust has done in our time. But whereas existential philosophers sided against God's existence, Gnostics argued only for His absence.

During the Renaissance, when the classical Greek and Roman texts—which had been protected during the so-called Dark Ages by Islamic scholars—were rediscovered, thinkers risked their lives to free themselves from religious dogma. Giordano Bruno (1548-1600) was burned at the stake for his seven years of resolute heresy. Espousing a pantheistic perspective in which God, life, and mind were part of an ever-changing universe, Bruno even thought distant worlds might harbor intelligent beings. The same Christian view that Bruno defied holds firm today: God is as superior to the universe as mind is to matter, or soul is to body. Flesh, a necessary evil, is unclean; only spirit is pure.

Cartesian License

At the dawn of modern science, French Catholic mathematician René Descartes (1596-1650) posited a fateful split between *res extensa*, material reality, and *res cogitans*, thinking reality. Only humans, Descartes argued, partake of God to the extent that they have souls. Even animals, though they seem to feel pain, are soul-less machines: "We are so accustomed to persuade ourselves that the brute beasts feel as we do that it is difficult for us to rid ourselves of this opinion. But if we were as accustomed to seeing automata which imitate perfectly all those of our actions which they can imitate, and to taking them for automata only, we should have no doubt at all that the irrational animals are automatons."[3]

On the authority of Descartes, live animals were

nailed to boards without remorse to illustrate facts of anatomy and physiology. Nevertheless, Descartes' presentation of the universe as a vast mechanism did serve to open up the cosmos for scientific investigation. Unfeeling nature could be analyzed with no fear of trespass. Nature, a vast lifeless mechanism, could be dismantled and manipulated, experimented upon with impunity. Man became the final earthly refuge of divine presence.

By splitting reality into human consciousness and an unfeeling, objective, "extensive" world that could be measured mathematically, Descartes paved the way for a scientific investigation of nature constructed according to the mathematical laws of God.

"God sets up laws in nature just as a king sets up laws in his kingdom,"[4] wrote Descartes. A kind of Cartesian license gave precedence to matter over form, body over soul, outward spatially extended nature over inner awareness. Matter, body, and nature could — unlike thought or feeling — be quantified, examined, and ultimately understood by mathematical physics.

This Cartesian license permitted the human intellect, through science, to enter a thousand different realms, from the very small to the very large, and even the invisible. The blueprint underlying the great mechanism of the cosmos was thought decipherable. Optical instruments were focused on snow flakes and peppercorns, or pointed at the pockmarked whiteness of the side-lit moon. Atoms were investigated by chemical combination and physical acceleration. X-rays imaged bones. Radioactive elements clocked the internal metabolism of the human body. Engineers even appropriated the seemingly God-given power to fly.

Flashing the Cartesian permit to practice science did yield results. Investigators returned *The Bible* and the classic books to their dusty shelves. Instead they read Nature, "written," as Galileo Galilei (1564-1642) had put it (even before Descartes), "in a great book which is always open before our eyes."[5] Galileo paid dearly for his scientific leanings. As quantitative mechanist, measurer of falling bodies, discoverer of the moons of Jupiter and the rotation of the sun, Galileo cleared the trail for curious successors. As defier of potent philosophers and Christian theologians, Galileo provoked the ire of Church authorities. Although he was not, like Bruno, burnt at the stake, he was, at age 58, brought before the Inquisition and charged with heresy. Galileo recanted his earlier claims so at variance with official Church doctrine and "admitted" that Earth is the center of the universe. Warned against further heresy, he was condemned to three years of weekly psalm recitations. He became a prisoner in his own country home. Galileo's immensely popular masterpiece, *Dialogue of the Two Chief World Systems*, was banned — until 1838. Pope Urban VIII, believing that he had been mocked as spokesman of the Church's cosmological views in Galileo's character "Simplicio," began the censorship.

If Galileo had worked under Cartesian permit he would have fared better. The devout Descartes abandoned work on a book manuscript in which he was putting forth similar views, when in 1633 he learned of Galileo's condemnation. Anxious to conciliate religion and science, Descartes gave great impetus to modern philosophy by doubting everything but the existence of his own doubting mind. The body was entirely mechanical, he held, but connected to the mind through the pineal gland, a pea-sized structure known at the time only in human brains. The pineal gland acted, Descartes believed, as a valve through which God was connected to the human soul.

To this day the Cartesian permit rallies scientists to study a universe that is wide open for investigation, but in the "fine print" is found the exception: the conscious human soul—which, in Descartes' time was unquestionably made in God's image. Moreover, the Cartesian permit still contains in the fine print this assumption: the universe is mechanical and set up according to immutable laws. Neither the exception nor the assumption is science. At the very heart of the Cartesian philosophy are thus metaphysical presuppositions, springing from the culture that gave rise to science.

Ultimately—in our very abbreviated story—the Cartesian license proves to be a kind of forgery. After three centuries of implicit renewal, the license is still accepted even though the fine print, erased or ignored, is no longer visible at any magnification. Yet this fine print was not incidental. It was the raison d'être, the rational basis authorizing scientists following the spirit of Descartes to proceed with their work and to receive the blessings of society, if not always the Church. The Cartesian view of cosmos as machine is at the very root of the practice of science.

Entering the Forbidden Realm

While Descartes cogitated, Europe remained under the rule of royalty. The King and the Lord, representing the power and order of God, reigned supreme. But science soon entered the forbidden realm of humankind, the one place it was not supposed to go. Scientific revelation of mechanism, part of the new audacity of inquiry, helped unsettle European monarchy. If the universe made by God is a giant automaton that works itself, why should people obey any King or Lord whose power, God-given in the feudal system of medieval Christianity, no longer derived from heavenly decree? The high-born Frenchman Donatien Alphonse François Sade keenly felt the vanishing basis for morality. If Nature was a self-perpetuating machine and no longer a purveyor of divine authority, then it did

not matter what he, as the infamous marquis de Sade, did or wrote.

In 1776 the British colonists in North America broke free from transatlantic rule. Independence from the burdens of taxes and royalty was proclaimed. In 1789 the French Revolution deposed the king and stripped the lords and ladies of their powers. Irreverent Voltaire claimed that if God did not exist it would be necessary to invent him. (A century later German philosopher Friedrich Nietzsche would declare God dead.) England, too, was struck by the revolutionary spirit of the time, but in moderation. Retaining their king and queen, the English perceived themselves a bastion of order in a world gone mad.

Enter Charles Darwin. In 1859 his *Origin of Species* was published, announcing to the world the scientifically derived inference that man had not been created by God, but had evolved from mere animals through "natural selection." Darwin's later books, *Descent of Man* (1871) and *Expression of Emotions* (1872), explored the then-startling thesis that humans and apes evolved from ancient apes. Darwin documented, without any explicit anti-Christian statement, that neither humans nor ancestral apes were created by God. The Great Chain of Being—the line of holiness coming down from God through spiritual angels to humankind and thence to the rest of mechanical creation— was turned topsy-turvy. The cosmic apple cart was upset. No longer, Darwin insinuated, was man excluded from connection with nature. Even the perceiving mind, describing itself, evolved from mechanical laws of random variation and natural selection. Materialism was victorious. As in some maudlin Disney animation, the last sparkle of fairy stuff disappeared.

Western thought thus suffered a metaphysical reversal. Once, before the exploits of Bruno and Galileo, Descartes and Newton, and Darwin,

11
The so-called "Grand Design" spiral galaxy, M-100, as taken from the Hubble Space Telescope on December 31, 1993. Our ancestors inhabited a world of living spirits from vengeful storm clouds to serenely shining stars; today science pictures the universe as an inanimate morass of chemical elements and compounds, some of which, under certain conditions, conjoin to form the self-repeating structures of life.

everything had been alive, except for the natural magic trick of death; now in the scientific-mechanistic world everything was inanimate, dead, except for the scientific puzzle of life. [PLATE 11]

We all are interested in life because we know it from the inside as something more than mechanical, automatic, determined responses to preordained stimuli. We think, act, choose. We—and it would be a conceit to exclude other organic beings—are not Newtonian machines.

Moreover, we are not objective outsiders. In physics, Werner Heisenberg's uncertainty principle limits what is measurable. In mathematics, Kurt Gödel's incompleteness theorem warns that every mathematical system, if complete, cannot be consistent and, if consistent, cannot be complete, since to define it axioms are needed from outside the system. Such scientific uncertainty also impedes any search to define life. On the one hand, a final definition of life by life may be like kissing your elbow or rolling your eyes to see your own optic nerve: impossible. On the other hand, enlightened by a knowledge of history and science's astounding success at investigating what life is, we seem closer than ever to a deeper understanding of life in its cosmic and cultural context.

In the flush of this exhilarating material success scientists tend to gloss the distinction between life and nonlife, pointing to the chemical continuities. Life-as-a-whole is like other vast subjects: nationalism, culture, politics, or anything else not easily defined, manipulated, or described. Even biologists may be snide, dismissing relevant discussion as "just philosophy." But science, like anything else, has a context. And that context is partly metaphysics, great, often-unstated categories of thought, perhaps cultural, perhaps inherited (the distinction is itself metaphysical!) that go beyond science proper. No one escapes metaphysics; to understand life, as science, it is necessary to understand its cultural context.

"Metaphysics," introduced by Hellenistic scholars and referring to certain untitled texts by Aristotle, comes from the Greek *ta meta ta physika biblia*, which literally means "the books after ('meta') the books on nature." The original use of the prefix "meta," by early editors such as Andronicus of Rhodes, may not have referred to any transcendental interpretation of ultimate reality, but only to the mundane position of the book on the table where "Metaphysics" was stacked on top of "Physics." Beginning with the work of Immanuel Kant, metaphysics has come to refer to speculations on questions not answerable by direct observation or experiment. Metaphysics, as a web of ideas in which we are caught, need not give rise to futility. It is fascinating to try and tease apart the strands of the culturally inherited, linguistically reinforced concepts that guide even our most seemingly original thoughts. An explanation of metaphysics may not lead to absolute truth but it certainly shouldn't be anathema to open, scientific minds.

Cosmic Wiggles

"A living body," wrote Alan Watts, "is not a fixed thing but a flowing event." Watts, the Anglo-American popularizer of eastern philosophy, drew from science, as well, in his quest for the meaning of life. He likened life to "a flame or a whirlpool":

> the shape alone is stable. The substance is a stream of energy going in at one end and out at the other. Life's purpose to maintain and perpetuate itself is understandable as a physico-chemical phenomenon studied by the science of thermodynamics. We are temporarily identifiable wiggles in a stream that enters us in the form of light, heat, air, water, milk.... It goes out as gas and excrement—and also as semen, babies, talk, politics, war, poetry and music."[6]

Thermodynamic systems lose heat to the universe as they convert energy from one form to another. Living matter frees itself from ordinary matter only by perpetually basking in the sun. Confronted with dissolution and destruction, life suffers a permanent death threat. Life is not merely matter, but matter energized, matter organized, matter with a glorious and peculiar built-in history. Life as matter with needs inseparable from its history must maintain and perpetuate itself, swim or sink. The most glorious organic being may indeed be nothing but "temporarily identifiable wiggles," but for millions of years as life has been racing away from disorder autopoietic beings have concerned themselves with themselves, becoming ever more sensitive, ever more future oriented, and ever more focused on what might bring harm to the delicate wave of their matter-surfing form. From a thermodynamic, autopoietic perspective, the basest act of reproduction and the most elegant aesthetic appreciation derive from a common source and ultimately serve the same purpose: to preserve vivified matter in the face of adversity and a universal tendency toward disorder.

Dutch-Jewish philosopher Baruch Spinoza (1632-1677) portrayed matter and energy as the fundamental nature of a universe which was itself alive. The great German writer and naturalist Johann Wolfgang von Goethe (1749-1832), author of *Faust*, argued for a poetic biology. He thought matter does not operate without spirit, nor does spirit exist without matter. Although he was pre-Darwinian and his theories are now obsolete,

Goethe wrote ably on science. In one passage he plucks from human activity what might be called its autopoietic essence:

> Why are the people thus busily moving? For food they are seeking,
>
> Children they fain would beget, feeding them well as they can.
>
> Traveler, mark this well, and, when thou art home, do thou likewise!
>
> More can no mortal effect, work with what ardor he will.[7]

The German biologist Ernst Haeckel (1834-1919), inventor of the word "ecology," promoted the idea that the activity of the human psyche is an offshoot of physiology: "We hold with Goethe that matter cannot exist...without spirit....We adhere firmly to the pure, unequivocal monism of Spinoza: Matter, or infinitely extended substance, and Spirit (or Energy), or sensitive and thinking substance, are the two fundamental attributes, or principal properties, of the all-embracing essence of the world, the universal substance."[8]

The Meaning of Evolution

Ernst Haeckel was Darwin's translator and greatest advocate in the German tongue, but he pushed Darwinism further than its inventor had been willing to carry it. The soul, Haeckel claimed, resided in the cell, immortality was a metaphysical sham, life had no purpose other than itself, and beings were not spiritual but material in nature. "Humanity," he declared, "is but a transitory phase of evolution of an eternal substance, a particular phenomenal form of matter and energy, the true proportion of which we soon perceive when we set it on the background of infinite space and eternal time."[9]

Such views infuriated traditional religious sensibilities, including those of Alfred Russel Wallace (1823-1913). An English naturalist, Wallace developed his own theory of evolution by natural selection that was uncannily similar to that of Darwin. Darwin's and Wallace's short papers on natural selection were published together in the same issue of the *Journal of the Proceedings of the Linnean Society of London, Zoology*. Wallace, who frequented seances, reviled Haeckel's notion of matter as eternal and alive, and he rejected Haeckel's denial of a spirit world. He sneered that *The Riddle of the Universe*, the title of one of Haeckel's most influential and popular books, had not been solved, least of all by Haeckel.

Even before Darwin, German philosopher Immanuel Kant (1724-1804) noted that skeletal and other similarities pointed to blood ties, a common parentage for all life. Kant ceded that all life could have arisen through some mechanical process similar to that by which nature produces crystals, but he judged it would be absurd to hope for "a Newton" who could make comprehensible even the growth of a single blade of grass by mechanical theory alone. Haeckel proposed Darwin as the very "Newton" Kant had believed impossible.

By projecting Earth history millions of years beyond the six thousand years allotted in the "Book of Genesis," James Hutton (1726-1797) founded modern geology. Hutton, son of a Scottish merchant, distinguished rocks laid down as sediment from those brought forth in molten form through volcanos. He observed erosion by wind and water and deduced the production of rainfall from cooling air masses that could no longer contain their moisture. Older sediments were deposited prior to more recent ones. Hutton's "law of superposition" led to Charles Lyell's (1799-1875) statement of the "law of uniformitarianism," the suggestion that only those geological forces observable in the present need be invoked to account for structures made and sediments accumulated in the past. But Hutton's extrapolation that Earth must be very old was controversial. Conservative England, threatened by the wild and godless French Revolution, was not ready to accept an Earth older than that which could be ascertained by summing up all the begats mentioned in *The Bible*.

Nonetheless, Scottish geologist Charles Lyell approved Hutton and argued that time was far vaster than previously thought in his multivolume book, *The Principles of Geology*—which did for that field what Darwin's opus later did for zoology and botany. Lyell was also far ahead of his time in taking a global ecological perspective reminiscent of Gaia theory today; he called attention to "the powers of vitality on the state of the earth's surface."[10] Darwin read Lyell during his voyage on the *Beagle* and adopted the Lyellian worldview. Decades later Lyell, in turn, embraced the Darwinian worldview. In 1863 he published *The Antiquity of Man*, which suggests, before Darwin had made the extension, that evolution applied to all humankind.

Meanwhile on the Continent, Berlin naturalist Christian Gottfried Ehrenberg (1795-1876) was putting the life back into biology. Returning from an ill-fated expedition to Egypt, of which he was perhaps the sole survivor, he focused on the transition between life and nonlife. In the expedition to Egypt (1820) and a later one to Siberia (1829) Ehrenberg documented the unseen world of microbes that fertilize the oceans and soils. Through his journeys Ehrenberg came to know Friedrich Wilhelm Heinrich Alexander, the Baron von Humboldt (1769-1859). Humboldt, widely regarded as the greatest German naturalist of his time, had collected more than sixty thousand plant specimens during his travels around the world. He had visited American president Thomas Jefferson and was described as a scientific "Napoleon." In his seventies Humboldt began to compile *Kosmos*, his grand attempt to map and explain the entire universe. "Certainly," wrote Isaac Asimov, "no man before him, with so active a mind, had seen so much of the world, and no man before him was so well equipped to write such a book.... It was a florid production, rather overblown, but it is one of the remarkable books in scientific history and was the first reasonably accurate encyclopedia of geography and geology."[11]

In *Kosmos*, Humboldt shares Ehrenberg's discovery of life's global sweep. "The universality of life is so profusely distributed," waxes Humboldt,

> that the smaller Infusoria [ciliates and other protists] live as parasites on the larger, and are themselves inhabited by others.... The strong and beneficial influence exercised on the feeling of mankind by the consideration of the diffusion of life throughout the realms of nature is common to every zone, but the impression thus produced is most powerful in the equatorial regions, in the land of palms, bamboos, and arborescent ferns, where the ground rises from the shore of seas rich in mollusca and corals to the limits of perpetual snow. The local distribution of plants embraces almost all heights and depths. Organic forms not only descend into the interior of the earth, where the industry of the miner has laid open extensive excavations and sprung deep shafts, but I have also found snow-white stalactitic columns encircled by the delicate web of an *Usnea* [old man's beard lichen], in caves where meteoric water could alone penetrate through fissures.... [Organisms flourish on the summits of the] Andes, at an elevation of more than 15,000 feet. Thermal springs contain small insects (*Hydroporus thermalis*), *Gallionellae* [iron bacterial masses], *Oscillatoria*, and *Confervae* [an old name for a

miscellany of green algae], while their waters bathe the root-fibers of phanerogamic [cone- and flower-bearing] plants.[12]

Humboldt died the same year Darwin published *The Origin of Species*.

Until very recently, with publication of the work of Schrödinger's legacy, observations made by Humboldt and Ehrenberg on the microbial world and many other late nineteenth century discoveries were not brought together in an evolutionary context. The fertilization of sperm by egg (embryo formation), inheritance factors of garden peas (Mendelian genetics), mucoid substances in the pus of soldier's wounds (nucleic acids, DNA and RNA), and visualization of chromosomes were some of the revelations made last century which, in geneticist Theodosius Dobzhansky's words, only "make sense…in the light of evolution."[13]

Although theories of evolution had been in the air for a half century and more, Darwin's methodical purposefulness, his diplomacy of prose, and his presentation as an Englishman of a mechanical theory during a time when Isaac Newton's theory of gravity was the last word in science all helped make the appearance of his book an epic event. As one woman of society wryly remarked on hearing the news of her less-than-noble origins, "Let us hope it is not true. But if it is, let us hope it does not become generally known."

Since *The Origin*, the idea of evolution has become increasingly accepted—overwhelmingly by scientists and respectably by the public (particularly the educated public). But it has also been abused. For example, in a popular illustration Haeckel depicted the summit of evolution as a nude but demure Germanic woman at the top of his evolutionary tree. Haeckel's error was not so much in his Germanic bias (or his choice of the female sex) but in his choice of any human at all. This is because all extant species are equally evolved. All living beings, from bacterial speck to congressional committee member, evolved from the ancient common ancestor which evolved autopoiesis and thus became the first living cell. The fact of survival itself proves "superiority," as all are descended from the same metabolizing Ur-form. The gentle living explosion, in a circuitous 4,000-million-year path to the present, has produced us all. In a sense then, the Vedic intuition that individual awareness is illusory and that each of us belongs to a single primal ground—Brahman—may be accurate: we share a common heritage, not only of chemistry but of consciousness, of the need to survive in a cosmos whose matter we share but which is itself indifferent to our living and self-concern.

Vernadsky's Biosphere

Given the limited legacy of metaphysical dualism (mind/body, spirit/matter, life/nonlife), it may not be surprising that two of the most profound rethinkers of life and its environment in this century share a biospheric perspective yet have diametrically opposed views. But whereas Russian scientist Vladimir Ivanovich Vernadsky (1863-1945) described organisms as he would minerals—calling them "living matter"—English scientist James E. Lovelock describes Earth's surface, including rocks and air, as alive.

Vernadsky portrayed living matter as a geological force—indeed, the greatest of all geological forces. Life moves and transforms matter across oceans and continents. Life, as flying phosphorus-rich gulls, racing schools of mackerel, and sediment-churning polychaete worms, moves and chemically transforms the planet's surface. Moreover, life is now known to be largely responsible for the unusual character of Earth's oxygen-rich and carbon dioxide-poor atmosphere.

Like Ehrenberg and Humboldt before him, Vernadsky showed what he called the "everywhereness of life"—living matter's almost total penetration into, and consequent involvement in, seemingly inanimate processes of rock, water, and wind. Others spoke of an animal, vegetable, and mineral kingdom; Vernadsky analyzed geological phenomena without preconceived notions of what was and was not alive. Perceiving life not as life but as "living matter," he was free to broaden its study beyond that of biology or any other traditional discipline. What struck him most was that the material of Earth's crust has been packaged into myriad moving beings whose reproduction and growth build and break down matter on a global scale. People, for example, redistribute and concentrate oxygen, hydrogen, nitrogen, carbon, sulfur, phosphorus, and other elements of Earth's crust into two-legged, upright forms that have an amazing propensity to wander across, dig into, and in countless other ways alter Earth's surface. We are walking, talking minerals.

Vernadsky contrasted gravity, which pulls material vertically toward the center of Earth, with life—growing, running, swimming, and flying. Life, challenging gravity, moves matter horizontally across the surface. Vernadsky detailed the structure and distribution of aluminosilicates in Earth's crust and was the first to recognize the importance to geological change of heat released from radioactivity.

But even a resolute materialist like Vernadsky found a place for mind. In Vernadsky's view a special thinking layer of organized matter growing and changing Earth's surface is associated with humans and technology. To describe it, he adopted the term *noosphere*, from Greek *noos*, mind. The term had been coined by Edouard Le Roy, philosopher Henri Bergson's successor at the College de France. Vernadsky and Le Roy met in Paris for intellectual discussions in the 1920s, along with Pierre Teilhard de Chardin, the French paleontologist and Jesuit priest whose writings would later bring the idea of noosphere—a conscious layer of life—to a wide

audience. Teilhard's and Vernadsky's use of the term noosphere, like their slants on evolution in general, differed. For Teilhard the noosphere was the "human" planetary layer forming "outside and above the biosphere," while for Vernadsky the noosphere referred to humanity and technology as an integral part of the planetary biosphere.

Vernadsky distinguished himself from other theorizers by his staunch refusal to erect a special category for life. In retrospect we can see the value of his stance; because life has indeed become a category, theorists of life have managed to reify—to make a thing out of—something that is not a thing at all. Vernadsky's referring to life as "living matter" was no mere rhetorical ploy. In one deft verbal stroke Vernadsky cut loose centuries of mystic clutter attached to the word "life." Making every attempt to consider life part of other physical processes, his use of the gerund "living" stressed that life was less a thing and more a happening, a process. Organisms for Vernadsky are special, distributed forms of the common mineral, water. Animated water, life in all its wetness, displays a power of movement exceeding that of limestone, silicate and even air. It shapes Earth's surface. Emphasizing the continuity of watery life and rocks, such as that evident in coal or fossil limestone reefs, Vernadsky noted how these apparently inert strata are "traces of bygone biospheres."[14]

Austrian geologist Edward Seuss had coined the word "biosphere," but Vernadsky brought it into currency. Just as the sphere of rock is a lithosphere, and that of air an atmosphere, so the sphere where life exists is a "biosphere." In his 1926 book, *The Biosphere*, Vernadsky showed how Earth's surface was an ordered transformation of the energies of the sun. "The biosphere," wrote Vernadsky, "is as much, or even more, the creation of the Sun as it is a manifestation of Earthly processes. Ancient religious intuitions that regarded Earth's creatures, especially humans, as 'children of the Sun' were much nearer

NASA x-ray photograph of the Sun. According to Russian scientist Vladimir Vernadsky (1863-1945) life on Earth is a material system in which stellar energies become transformed into living ones. Life is not only a global but a planetary/solar system phenomenon.

the truth than those that looked upon them…as a mere ephemeral creation, a blind and accidental product of matter and Earth forces…We may regard living matter in its entirety, then as the peculiar and unique domain for the accumulation and transformation of the luminous energy of the Sun."[15]

Remarkably, Vernadsky dismantled the rigid boundary between living organisms and a nonliving environment, depicting life globally before a single satellite had returned photographs of Earth from orbit. Indeed, Vernadsky did for space what Darwin had done for time: as Darwin showed all life descended from a remote ancestor, so Vernadsky showed all life inhabited a materially unified place, the biosphere. Life was a single entity, transforming to earthly matter the cosmic energies of the sun. [PLATE 12] Vernadsky portrayed life as a global phenomenon in which the sun's energy was transformed. Emphasizing photosynthetic growth of red and green bacteria, algae, and plants, he saw these expressions of living matter as the "green fire" whose expansion, fed by the sun, pressured other beings into becoming more complex and more dispersed.

Vernadsky set forth two laws. Over time, he claimed, more and more chemical elements became involved in the cycles of life. Second, the rate of migration of atoms in the environment has increased with time. A flock of migrating geese was to Vernadsky a biospheric transport system for nitrogen. Locust swarms, recorded in *The Bible*, attested to massive changes in the distribution of carbon, phosphorus, sulfur, and other biologically important chemicals two thousand years ago. As dams, factories, mines, machine construction, utilities, trains, planes, global communications, and entertainment systems have appeared, more chemical elements than ever have become organized into functioning parts of autopoietic systems. Technology, from a Vernadskian perspective, is very much a part of nature. The former calf muscle severed into brochette cubes and the pine tree trunk into lumber pass through the hands of workers and the chutes of machines to emerge transformed into shish kebob and floor. The plastics and metals incorporated in industry belong to an ancient process of life co-opting new materials for a surface geological flow that becomes ever more rapid. And, with the fleeting synthesis in physicists' laboratories of radioactive isotopes, the noosphere begins to direct and organize atoms that have never before existed on Earth.

Lovelock's Gaia

As Vernadsky disrupted the mind/matter split through a consideration of living matter on a global scale, so James E. Lovelock upsets metaphysical dualism by an opposite stratagem—considering Earth alive. Vernadsky examined life as matter within a receptive political and cultural climate—the official atheism of the former Soviet Union, aided by science's approval of materialism. By contrast, Lovelock—portraying the self-regulating biosphere, a huge and oddly spherical living body he calls "Gaia"—has been hampered by the subtle ideology of mechanism that pervades the scientific community. This means that Lovelock must not only show that Earth maintains itself as a living body, he must also overcome the prejudice that to call this "thing" alive is not science but poetic personification. Given these tensions, it is a testament to this world-class atmospheric chemist's ingenuity that his theory is taken as seriously as it is by active scientists.[16]

Atmospheric, astronomical, and oceanographic evidence attest that life manifests itself on a planetary scale. The steadiness of mean planetary temperature for the past three thousand million years, the 700-million-year maintenance of Earth's reactive atmosphere between high-oxygen levels of combustibility and low-oxygen levels of asphyxiation, and the apparently continuous removal of hazardous salts from oceans—all these point to mammal-like purposefulness in the organization of life as a whole. [PLATES 13 and 14]

This purposefulness, central to scientific Gaia theory, is a major sticking point for traditional biologists. How can a planet behave in a purposeful manner to maintain environmental conditions favorable to its living constituents? In mechanistic biology, complex self-regulation only evolves from natural selection that weeds out more poorly self-regulating individuals. This logic is flawed, however. According to it no original, self-maintaining cell could have ever evolved, because "purposeful," self-regulating behavior simply cannot arise in a population with only one member. A strict reading of Darwinism denies evolutionary capabilities to a population of one.

Plausible within the bounds of Darwinism or not, both planet, isolated by space, and cell, isolated by semipermeable membrane, are solar energy-requiring systems, continuous through time and space, that display self-maintaining behavior. The

"purposefulness" of Gaian self-maintenance derives from the living behavior of myriad organisms, mostly microbes, whose ubiquity Ehrenberg and Humboldt first established. Planetary physiology, far from having been produced *ex nihilo*, or by an outside God, is the holarchic outcome of ordinary living beings. It is the autopoiesis of the cell writ large.

Life cannot be understood ignoring the sentient observer. If not for mind, no one would care that life is a certain kind of sunlight-energized cosmic debris. But it is, and we do. To best understand life we need to see the long and winding road from animism, through dualism, to the limitations of mechanism. Physics, chemistry, and biology are distinct approaches to the same material phenomena. As German geomicrobiologist Wolfgang Krumbein puts it,

> The mineral and microbial mineral cycles as we view them today on the basis of experimental work have been envisaged as the unifying

concept of world and universe, creating the principle of the one living nature of Bruno and Spinoza....The basic approach of Bruno... is still alive and is evidenced in scientific and mathematic terms by non-Euclidian geometry, by the modern field theories and Einstein's relativity and gravity theories, as well as by Lovelock's "Gaia-hypothesis." Giordano Bruno deeply influenced Baruch Spinoza, Leibnitz, Kant, Goethe and Schelling. He still influences unitarian thought in science and philosophy... a textbook on microbial geochemistry...must come back to Bruno's original thoughts of "cyclic developments" rather than "creation and destiny" as revealed in the clerical Christian thoughts of his time which have so severely inhibited the development of science.[17]

Life can be returned to biology without compromising science. Mechanism gave science the authority to examine the realms of heaven and life once considered "off limits." But it also suggested the universe was more deterministic than it is, cutting into our sense of life and wonder. The Epicurean Roman philosopher Lucretius (95-55 B.C.), in his poem *De Rerum Natura* ("On the Nature of Things"), presents an evolutionary view of the universe denying a hereafter and arguing that everything, even the soul and gods, are made of atoms. In the same tradition, Bruno blended matter with energy, finite with infinite, world with God. In the modern era, by not speaking of life at all — but calling it "living matter"—Vernadsky offered us a chance to see life with fresh eyes. And, unlike monolithic Cartesian materialism, the Gaia perspective accommodates the enchantment we feel as living beings dwelling in a living world.

So, what is life?

Life is planetary exuberance, a solar phenomenon. It is the astronomically local transmutation of Earth's air, water, and sun into cells. It is an intricate pattern of growth and death, dispatch and retrenchment, transformation and decay. Life is the single expanding organization connected through Darwinian time to the first bacteria and through Vernadskian space to all citizens of the biosphere. Life as God and music and carbon and energy is a whirling nexus of growing, fusing, and dying beings. It is matter gone wild, capable of choosing its own direction in order to indefinitely forestall the inevitable moment of thermodynamic equilibrium — death. Life is also a question the universe poses to itself in the form of a human being.

What happened to living matter to make it so different? The answer is both scientific and historical. Life is its own inimitable history. From an everyday, uncontentious perspective "you" began in your mother's womb some nine months before whatever your age is. From a deeper, evolutionary perspective, however, "you" began with life's daring genesis — its secession, more than 4,000 million years ago, from the witches brew of the early Earth. In the next chapter we see how this brew, sometimes called the primeval soup, started percolating.

3
Once Upon the Planet

If a dirty undergarment is squeezed…a ferment drained from the garments and transformed by the smell of the grain, encrusts the wheat itself with its own skin and turns it into mice.…And, what is more remarkable, the mice from corn and undergarments are neither weanlings or sucklings nor premature but they jump out fully formed.

—Jean Baptiste van Helmont

For the humblest organism, the simplest bacterium, is already a coalition of enormous numbers of molecules. It is out of the question for all the pieces to have been formed independently in the primeval ocean, to meet by chance one fine day, and suddenly arrange themselves in such a complex system.

—François Jacob

It must be admitted that no one yet knows how life began.

—Stanley Miller and **Leslie Orgel**

Beginnings

On Earth some four thousand million years ago life generated as matter undertook a different direction. From the beginning, life satisfied its autopoietic imperative in a universe obeying thermodynamic laws. Bound and separated from the world by a border of its own making, life came together as oily droplets that increased their order. [PLATE 15] Other dissipative systems in nature use energy to increase order, but they last for only short periods of time. Moreover, a tornado risen on the plains doesn't go "whoops" as it wanders into a mountainous landscape that spells its doom; but even the simplest life form effectively does, actively responding to its surroundings to preserve and protect its form.

How matter in a bath of energy (or how energy in a brew of matter) first accomplished the feat of life is not known. No molecule by itself reproduces. Minimal life on Earth today is a system, a minute membrane-bounded sphere, a bacterial cell, requiring many interacting molecules. Some 2,000 to 5,000 genes make a similar number of proteins. Proteins and DNA mutually produce each other within the cell membrane that together they fabricate. Bearing a common biochemistry, all life probably dates to a single, perhaps (but not necessarily) improbable historical moment. The factors that led matter to its peculiar "fractionation point" where dissipative behavior became living behavior need only have happened once. Enclosed, perhaps even suddenly, by a membrane and with resources aplenty, the first living cells could afford to be somewhat aloof from external reality. Eventually, imperiled by its own profligacy and by the insensitivity of the substance from which it seceded—yet upon which it absolutely depended for sustenance—life was left to its own devices. As matter ostracized from itself, life had been abandoned by the world, yet the world had gone nowhere. There was no going back.

Once begun, reproducing systems proceeded

Cylinders of phospholipids produced through a hydration/dehydration process by David Deamer in his laboratory. As seen here through a light microscope at high magnification, they eventually break apart to form liposomes. Such precellular organic formations were commonplace in Archean times. The early Earth was like a giant laboratory in which countless experiments were occuring all the time. "God," jests organic chemist Cyril Ponnamperuma, "is an organic chemist."

rapidly away from their initial state, and today no vestiges of early life less complex than a bacterial cell remain. Bacteria are not half-hewn but fully living and evolved beings that have been thriving for more than 3,500 million years. The greatest chemical inventors in the history of Earth, they are not "just germs." Because of the conservative material nature of reproducing life, bacterial cells retain clues to the chemistry of Earth's surface as it existed in the remote past. Bacteria were the first green beings to grow on nothing but sun, water, and air. Still the only beings able to perform many metabolic tricks of which we animals and even plants are not capable, bacteria were the first to breathe oxygen and to swim. They are the virtuosi of the biosphere. They are also our relatives, which probably explains why we feel free to malign them.

Bacteria, never having gone extinct, continue to protect us as their populations grow prodigiously. They maintain soils for us and purify waters. Bacteria expel gases, filling their immediate environment with wastes noxious to the producers but alluring to other strains as they colonize ubiquitous niches—even glacial ice and boiling hot springs. Some build durable structures and crowd them with their communities; some make vinegar; others work metals such as iron, manganese, and even gold. Some sense the sun, swimming to bask in its rays, while others are photophobic. Still other bacteria sense and swim toward the closest magnetic pole. Many bacteria are poisoned by oxygen; others thrive in it. Some make spores remarkably resistant to heat, desiccation, or radiation. Bacteria come in a variety of colors, from snow-white *Beggiatoa* and yellow sulfur bacteria to red *Chromatium* and blue-green cyanos like *Spirulina*, *Nostoc*, or *Microcystis*. Bacteria, in short, are no more "germs" than the plants that feed, clothe, and house us are "weeds."

But how did the first bacterium originate? Again, no one knows. Bacteria are so sophisticated that they could have come from space. In the fifth century B.C. the Greek scientist Anaxagoras, friend of the playwright Euripides, invented "panspermia," the notion that life, dispersed as seeds throughout the universe, landed on Earth. Later, Swedish chemist Svanté Arrhenius (1859-1929), Nobel prize-winner for his ion theory that atoms in solution generate electrical charge, proposed that hardy bacterial spores were pushed by solar wind from star to star. Arguing that minute particles would rise to the upper atmosphere in volcanic eruptions, and that some of these particles, carrying bacterial spores, would reach the stratosphere—where electrical discharges would propel them into space—Arrhenius calculated that an Earth spore so launched today would reach Pluto in four months and arrive at Alpha Centauri, the nearest star, in just 7,000 years. (Spores discovered in peat and other deposits older than that have proved to be viable.) More recently, Francis Crick, codiscoverer of the structure of DNA, argues for "directed panspermia"—that intelligent extraterrestrials may have seeded Earth with the cosmic equivalent of a start-up kit for life.[1]

Might life have begun in some other star system and then have migrated (or been exported) to Earth? Perhaps, but such a view is less amenable to scientific investigation than the view that life originated right here on Earth. Moreover, if life started in outer space—say, on an Earthlike planet—study of how life arose would apply to its beginning anywhere. Indeed, Earth itself is suspended in space, so any way we look at it life came from space.

Hell on Earth

It is dawn, 4,600 million years ago. Earth is in the violent red throes of its beginnings, a gravitational implosion of molten rock and swirling metal. Superheated gases such as ammonia, hydrogen sulfide, and methane curl in an atmosphere zapped everywhere and at every moment by lightning. The very oceans hang unfallen, unrained—a sphere of steam eclipsing any semblance of sun. Beneath these vapors, thick with formaldehyde and cyanide (simple organic compounds that form naturally in space), the boiling crust of the surface pullulates with radioactivity and heat.

Meanwhile, the sun has ignited, flashing on with a burst of radiation so forceful it blows off whole planetary atmospheres, relegating hydrogen gas to the outer reaches of the solar system. There hydrogen collects around cold and massive Jupiter, Saturn, Uranus, and Neptune—for only these giant planets have gravity sufficient to retain their original allotment of this lightest of all elements. Everywhere, on all planets and their moons, meteors ranging in size from dust specks to small planetoids continue their bombardment. Smashing their way around the solar system, such space rubble brings water and carbon compounds along for the ride, thickening the brew that will feed early life on Earth.

One particularly huge intruder strikes Earth, flinging continent-sized masses into space. But it is slowed enough by the collision to become entrapped in Earth's orbit. After receiving a few more smaller smashes of its own, the cratered orb comes to resemble the white, sun-glinting Moon that enchants us today. But in that far-off era the scene would not have been serene. The young Earth spun so fast that daylight lasted a mere five hours. The atmosphere, lacking oxygen, would have supplied no breath and no blue vistas.

Such was Earth as it might have appeared in the Hadean eon 4,600-4,000 million years ago. Life may well have been around, however, by the late Hadean—once the molten surface of Earth had sufficiently cooled and no extraterrestrial impact was large enough to rearrange the entire crust at a single strike. This earliest violent period of Earth history, named after Hades, Greek hell and abode of the dead, is the first of four long eons. [CHART D, *following page*]

Fossils, whether stony tree trunks, traces of worm burrows, footprints along petrified shorelines, spores buried in lake sediments, or oily chemicals from decaying leaves, are evidence of past life. No fossils—not even a volcanic rock—survives from Earth's Hadean eon. The Hadean on Earth can only be inferred from measurement of far older matter from meteors and from the moon. Some of the oldest rocks dating from the following eon have, however, survived, and these, indeed, define the onset of the Archean (4,000-2,500 million years ago). A few "unmetamorphosed" Archean rocks—those not subjected to altering heat and pressure—retain traces of life. Australian rocks 3,485 million years old contain eleven or more types of recognizable fossil bacteria. The oldest rocks on Earth today thus contain vestiges of life. No one knows when life began, but life is at least as old as we could empirically know it to be.

Spontaneous Generation

In Greek myth, goddesses issue from sea shells and mortals can be turned into animals or trees. "In all the things of Nature there is something marvelous,"[2] wrote Aristotle, who, because he looked to the real world rather than Greek myths for knowledge, is recognized as the first biologist or naturalist of the western world. Nevertheless, Aristotle accepted as fact the (to us) mythlike notion that matter suddenly springs to life.

We now think organisms reproduce, but our ancestors imagined that life, through some sort of fathering principle, spontaneously generated: God produced Eve from Adam's rib, meat decayed into maggots, one thing became another. A certain perceptual logic of proximity and likeness suggested that decomposing vegetation brings forth insects and that fireflies could issue, as Aristotle taught, from glinting morning dew. Augustine (345-430) argued that just as God could bypass grapes and turn water into wine, so He could bypass parents. Animals would thus appear directly from *occulta*

Earth History Timeline

Here our "Human-centered" or distorted-scale timeline represents major groups of organisms and events in Earth history as people and books usually portray them: with no regard for chronological symmetry in a world consisting only of large animals and plants. We follow this with a true-scale timeline (beginning on page 55) which denies center stage to people and other mammals and does not short shrift the first 4,000 million years of our planet's history. While it is often thought that nothing of interest occurred on Earth until the origin of the skeletalized Cambrian marine animals some 570 million years ago, a true appreciation of history requires that the great early chapters of life's story not be dismissed. The true-scale timeline depicts some of the most significant events in the story of life before the evolution in the Phanerozoic Eon of familiar life forms.

Human-centered or Distorted Scale Timeline

(Millions of years ago)

0.005   Appearance of cities.

0.01   **Holocene epoch.**
Appearance of agricultural urban centers based on agricultural cultivation of grasses (wheat, rice, etc.).

0.05   Appearance of culturally modern, cave-painting humans.

0.20   Appearance of anatomically modern humans (*Homo sapiens*).

1.65   Quaternary Period.
Pleistocene epoch.
Homo erectus and *Homo neanderthalensis* in the Middle East, Africa, and Europe.

4   Appearance of human ancestors (*Australopithecus afarensis*).

5   **Pliocene epoch.**
Mediterranean basin entirely dry and major faunal "turnover event" in Africa.
Diversification of Miocene apes (*Proconsul*) including hominids (e.g. *Ramapithecus*, australopithecines).

23   Neogene Period.
Miocene epoch.
First widespread grasslands.

35   **Oligocene epoch.**

40   Appearance of angiosperm herbs and trees with fruit.

57   **Eocene epoch.**

65   CENOZOIC ERA.
Paleogene Period.
Paleocene epoch.
Beginning of mammal diversification, including spread of primate order.
Abundant and widely dispersed mammalian fossils belonging to extinct families and genera.
Second largest extinction in history of life (including non-avian dinosaurs.)

100   Opening of what will become the Atlantic Ocean.
Appearance of flowering plants (angiosperms) and primates (plesiadapiforms).

145   Cretaceous Period.

200   Appearance of hard silica tests of diatoms.

208   Jurassic Period.

245   MESOZOIC ERA.
Triassic Period.
Largest extinction event in history of life.
Beginning of breakup of Pangea continent.

290   Permian Period.
Formation of extensive salt deposits, indicating inland seas and coral reefs and possible biospheric control of ocean salinity.
Appearance of large amphibians; mammal-like reptiles; bird-like reptiles, including dinosaurs; and shrew-like mammals.
Radiolarians and other protists abound in marine sediments.

300   Extension of reef-building coral animals (coelenterates) and coralline (rhodophyte) algae.

323   Pennsylvanian Period.
Widespread large trees in swamps lead to coal forests.

362   Mississippian Period.
Widespread occurrence of fish and amphibian vertebrates in fossil record.

408   Devonian Period.
Appearance of armored fish and invertebrate marine animals.
Land extensively covered by first forests.
First appearance plants with seeds.

440   Silurian Period.
Appearance of terrestrial plants, Rhyniophytes with fungi in plant roots.
Beginning of widespread life on land.

500   Colonization of land surfaces by algae and insects.

510   Ordovician Period.
Appearance of first (jawless) fishes.

570   **PHANEROZOIC EON.**
PALEOZOIC ERA.
Cambrian Period.
Appearance in fossil record of Cambrian hard-bodied "animals" and "plants": trilobites, foraminifera, dinomastigotes, radiolarians, and red algae.

4,600   **HADEAN EON.**
Origins of Earth-Moon system and other solar system planets.

4,600   4,500   4,400   4,300

HADEAN EON

Origins Earth-Moon system and other solar system planets.

Oldest rocks dated by radioactivity from meteor (a chondrite from Canyon Diablo crater, Arizona).

Outgassing of volatiles from mantle to atmosphere.

Abundant impact cratering.

Oldest known mineral crystals (present-day Australia).

Possible existence of first continents.

---

semina, invisible seed. Around the year 1000, Cardinal Pietro Damiani insisted that birds bloom from fruits, and ducks emerge from sea shells. English scholar Alexander Neckam (1157-1217) specified that fir trees, exposed to sea salt, give rise to geese. The Flemish alchemist and physician Jan Baptiste van Helmont (1580-1644) shared his recipe for making mice from dirty underwear.

We moderns may laugh, but the notion of spontaneous generation made such sense at the time that few questioned it. "Since so little is required to make a being," agreed Descartes, "it is certainly not surprising that so many animals, worms, and insects form spontaneously before our eyes in all putrefying substances."[3] Aristotle had taught that the heat of the male seed animated and formed the cooler matter carried in the woman's womb. Lacking sufficient male heat, a woman miscarried or gave birth to a limbless infant. Heat might bypass seed altogether and directly generate worms, bats, snakes, crickets, or other vermin from meat or filth. Alchemists used heat to try to synthesize gold. In a patrilineal, male-dominated Europe women were like potter's kilns in which the act of fathering came to fruition; the female supplied only matter and not the essence of living form. Even Newton suggested that plants might spring forth from the coruscation of cometary tails. Nor did the invention of the microscope sweep away the old idea. Many believed that the "animalcules" Leeuwenhoek had discovered in plant fluids, ditch water, and saliva emerged directly from these fluids, just as veal—left to its own devices—was thought to generate flies.

Ironically, the notion of spontaneous generation was at first threatened as much by the idea of fixed species as it was by countervailing observations.

Species were recorded as fixed categories. The works of Swedish botanist Carolus Linnaeus (1707-1778), the founder of modern taxonomy who gave the name *Homo sapiens* to the human body (but not to the soul), and those of French anatomist Georges Cuvier (1769-1832), who extended the Linnaean classification to fossils, made the notion of spontaneous generation more difficult to accept. For Linnaeus, fixed species were distinct and separate forms created by an omnipotent God. Cuvier believed fossils evidence of past life, in particular of catastrophic floods, at least one of which was recorded in *The Bible.*

It thus came to be believed that an all-powerful God, once and for all, had created all Earth's "creatures." Indeed, Swiss naturalist Charles Bonnet (1720-1793) ruled out spontaneous generation because it was superfluous to his theory of *emboîtement*—that the original female of each species had been created, as in a set of Russian dolls, with the germ cells of all future generations already inside her. Half-blind Bonnet had discovered the all-female reproduction system of certain insects, the parthenogenesis of aphids, a fact which helped him argue against "evolution"—the word he used to refer to the belief of those foolhardy enough to believe in the wanton notion of species transformation.

Spontaneous generation was adhered to even after the Florentine physician and poet Francesco Redi (1626-1697) performed his diligent experiments disproving spontaneous generation. Redi placed a variety of meats—a snake, some fish, and a slice of veal—in sealed jars. Another set of jars were left open. Redi's experiment was a clear success. In his "observations on the generation of insects," he recorded that he "began to believe that all worms found in meat were derived from flies, and not putrefaction."[4] Redi, in other words, developed a theory of maggots. Having seen flies hovering around and entering the open (but not the closed) jars, he confirmed his suspicion that the sealed meats, despite their putrid stench, did not become "wormy." In phase two of the experiment he covered meat with a cloth that prevented flies from laying eggs. No vermin appeared. He concluded that, "Earth, after having brought forth the first plants and animals by order of the Supreme and Omnipotent Creator, has never since produced any kind of plants or animals, either perfect or imperfect; and everything which we know in past or present time she had produced, came solely…from seeds of the plants or animals themselves, which thus, through means of their own, preserve the species."[5]

Scientists are said to abandon theories as soon as they are contradicted by experiment. In fact, many do the reverse, ignoring awkward experimental evidence in an effort to save appearances. Nowhere, to paraphrase Mark Twain, is so much derived from so little as in the production of scientific theory from scientific fact. A century after Redi's experiment, the English naturalist and Roman Catholic priest John Tuberville Needham (1713-1781) collaborated with the early evolutionist Georges Louis Leclerc Buffon (1707-1788). Buffon, as keeper of the Jardin du Roi, the French royal botanical gardens, was author of the forty-four volume *Natural History*, read by many of the educated class, including Erasmus Darwin, Charles's grandfather. Together, Needham and Buffon performed an experiment designed to determine whether spontaneous generation applied to all of life. Boiling

mutton broth, they carefully sealed it in a jar. Opening it a few days later, they saw copious growth, suggesting to them that spontaneous generation did apply to microbial life. Although absolutely misleading—because they failed to kill the boil-proof microbes—the experiment ironically confirmed Buffon's essentially modern notion that "organic molecules" could under certain conditions combine to produce microorganisms.

In 1768 Italian biologist Lazzaro Spallanzani (1729-1799) demonstrated that his illustrious predecessors Buffon and Needham had neglected to boil the broth sufficiently. Still, Spallanzani's tests did not satisfy Ernst Haeckel who believed prolonged heat destroyed a "vital principle" in the air. Not until French chemist Louis Pasteur (1822-1895) exposed boiled meat extract to air by means of a flask, whose long neck was bent down and then up, were vitalists defeated. Air, but not bacteria, yeasts, or any other sort of life, could rise against gravity to enter the zigzag passageway to the life-supporting broth. As soon as the glass was broken, and microscopic life could enter, growth on the broth began. No other explanation held: life came only from previous life that was begotten by still earlier life. And yet, the work of Pasteur, proving that life comes only from previous life, strongly suggested that only God could have created life in the Beginning.

Origins of Life

In 1871 Darwin mused that one "could conceive in some warm little pond, with all sorts of ammonia and phosphoric salts, light, heat, electricity, etc." a chemically formed "protein compound…ready to undergo still more complex changes."[6] To trace life

First appearance of banded iron formation (BIF), suggesting local sources of oxygen at sediment-water interfaces.

Barberton Mountain Land (present-day South Africa) and Pilbara Block (present-day Western Australia), containing fossil evidence for anoxygenic communities: microfossils, stromatolites, and chemical fossils.

Onverwacht Group and Warrawoona Group (present-day South Africa), containing abundant reduced carbon in shales, microfossils, and stromatolites, imply widespread occurrence of photosynthetic bacterial communities.

Earliest direct evidence of tectonic activity: granite of the Kaapvaal Craton from present-day South Africa.

Development of thickest (and therefore oldest) portions of continents.

Trace amounts of oxygen gas (O_2) in atmosphere and sediments.

back to matter was a logical extension of the idea that all species had evolved from a common ancestor. If species could evolve, what was to stop matter itself from evolving into life?

A young Russian biochemist, Alexander Ivanovich Oparin (1894-1981), published a book in 1929 entitled *The Origin of Life*. Oparin focused attention on specific ways in which chemicals might self-organize toward life. He described droplets growing by absorbing carbon compounds in a primeval soup. Theorizing an early hydrogen-rich atmosphere with gases such as methane and ammonia, and a solar source of energy, Oparin postulated that his "coacervates," or "semiliquid colloidal gels" would become increasingly dependent on their "own specific internal physico-chemical structure." Eventually,

> The internal structure of the droplet determined its ability to absorb with greater or less speed and to incorporate into itself organic substances dissolved in the surrounding water. This resulted in an increase in the size of the droplet, i.e., they acquired the power to grow.... A peculiar selective process had thus come into play which finally resulted in the origin of colloidal systems with a highly developed physico-chemical organization, namely, the simplest primary organisms.[7]

Because Oparin inhabited a nation (the former Soviet Union) that had been officially atheistic since 1917, he could theorize on this new version of spontaneous generation without confronting established religion.

In 1929 the British physiologist J. B. S. Haldane published an article making the point that reactive oxygen would have destroyed organic compounds; developing life, therefore, must have arisen in an oxygen-free atmosphere.[8] The work of Haldane and Oparin was an inspiration to "origins-of-life" experimenters from the United States, such as Stanley L. Miller, Sidney Fox, and Cyril Ponnamperuma. Nonetheless, Oparin was no more removed from his sociocultural milieu than were his predecessors; after World War II he declared Schrödinger's book *What is Life?* to be "ideologically dangerous" and he protested the new emphasis on genes, viruses, and nucleic acids, calling it "mechanistic." Yet Oparin, by imagining how life could have first evolved, revived the notion of spontaneous generation of life from nonlife.

In 1959 organic chemist Sidney Fox and his colleagues cooled water-free mixtures of amino acids to make "proteinoid microspheres." Resembling cocci bacteria, these microspheres would, under pressure, occasionally divide. Leslie Orgel of the Salk Institute in California discovered a DNA-like molecule (fifty nucleotides long) that formed spontaneously from simpler carbon compounds and lead salts. Five years later, ATP—the compound that is universally used by life to store energy—was produced by Carl Sagan, Ruth Mariner, and Cyril Ponnamperuma in a phosphorus-containing mixture of gases thought similar to Earth's early atmosphere. "It is, perhaps, ironic," writes University of Maryland chemist Ponnamperuma, "that we tell beginning students...about Pasteur's experiments as the triumph of reason over mysticism yet we are coming back to spontaneous generation, albeit in a more refined and scientific sense, namely, to chemical evolution."[9]

3,200	3,100	3,000	2,900	2,800

Continental tectonic activity—many small plates.

Formation of Fig Tree Group (present-day South Africa) of rocks that contain micro-fossils of reproducing cells.

Oldest evidence for life in present-day North America: Steep Rock, Ontario.

Widespread stromatolite reefs preserved at Steep Rock and Pongola Belt (present-day South Africa).

Diversification of bacteria—probably all major metabolic modes evolved by now (e.g., chemoautotrophy such as H_2, H_2S, NH_3, and CH_4 oxidation; oxygenic photosynthesis; reduction of iron and manganese oxides to metals.)

Gold deposited in paleoriver in present-day Witwaters-rand, South Africa, indicating bacterial-mediated gold precipitation in ancient estuaries.

Large continents formed from raised portion of the plates known as the "preCambrian shields."

The "abiotic production" of ATP was actually a continuation of work begun by Stanley L. Miller, a graduate student of Nobel laureate Harold Urey, at the University of Chicago in 1953. Miniaturizing what he thought was Earth's earliest ambience, Miller filled flasks with gases (imitation atmosphere) over the surface of sterilized water (imitation ocean). For a week he bombarded his glassware microcosm with a lightning-like electrical discharge. The result was the evolutionary biologist's version of the just-twitched limbs of Mary Shelley's Frankenstein monster. Alanine and glycine, two chemicals essential to living proteins, as well as many other compounds, had spontaneously appeared in the flasks. In the laboratory, cooking from scratch, scientists had thus repeated the prebiotic origin not quite of life but of the nutrients needed for self-maintenance—a sort of primeval food.

Miller's laboratory mock-up of the early planetary atmosphere contained hydrogen gases like those left over from the gravitational accretion of the sun: hydrogen (H_2), water vapor (H_2O), ammonia (NH_3), and methane (CH_4). The experiments showed in startling fashion that the chemicals of life do self-organize without conscious direction. Given favorable conditions—Miller's model of the early atmosphere was only a rough guess—organic compounds spontaneously form from simpler precursors. The undeniable conclusion was that at least the matter of life spontaneously generates.

Miller's experiments were repeated and modified by many enthusiastic chemists. Some used alternative sources of energy, such as ultraviolet radiation and heat. Akiva Bar-Nun, for example, generated "sonic booms" in the laboratory; he showed that even energetic sound waves make protein components from atmospheric gases. Adenine, cytosine, guanine, thymine, and uracil—the five nucleic acid bases that strung together make DNA or RNA molecules—all have been synthesized in "prebiotic chemistry" experiments.

Of the six kinds of atoms crucial to life on Earth—carbon, nitrogen, hydrogen, oxygen, sulfur, and phosphorus—all have been detected in space. [PLATE 16] Hydrogen, the most common element in DNA, RNA, proteins, fats, and other compounds created by life, is also the most common in the universe. Ammonia (NH_3) was discovered in interstellar space in 1968. H_3C_2N, cyanoacetylene, was detected in 1970. Alcohol (CH_3CH_2OH) abounds in the constellation Orion. Other compounds found both in space and in living things include water, acetylene, formaldehyde, cyanide, methanol (wood alcohol), and five-atom formic acid, the clear fluid secreted by agitated ants. "God," declared Ponnamperuma, "is an organic chemist."

Ponnamperuma thinks our planet may have been "knee-deep" in polyaminomaleonitryl, an organic compound whose combinations could have inaugurated the later world of cells. Polyaminomaleonitryl is a polymer, a large molecule made of repeated links of HCN, hydrogen cyanide. HCN, a simple three-atom compound, has been detected on Titan, Saturn's sixth (and biggest) moon. A precursor to other biochemicals, including adenine and guanine in the nucleotide bases of RNA and DNA, HCN may be a key ingredient in the cosmic recipe for life. Ponnamperuma casts it as the "God molecule." The polymer comes in a range of colors, including the reds and browns characteristic of

2,700	2,600	2,500	2,400	2,300

PROTEROZOIC EON

Stromatolites abundant and cosmopolitan on ancient continents in parts of present-day Africa, North and South America, Australia, and Asia.

End of major crust-forming period.

Geologically modern processes begin: Oxygen gas (O_2) begins to seasonally accumulate; banded iron formations (BIFs) conspicuous and abundant; extensive huge lakes or oceans; carbonate platforms, indicating biogenic reef-like structures made by bacterial communities in marine settings.

First super continent (prePangea).

Beginning of worldwide age of BIFs: 90% of Earth's current mineable iron deposits in present-day southern Africa, Brazil, Central America, western Ontario, northern Michigan, and Minnesota formed between 2,400 and 1,800 mya.

Continued expansion of carbonate reef-like platforms and BIFs.

16
Triton, a photographic montage composed from a dozen different images of this moon of Neptune. Many of the outer planets and their moons possess water ice and carbon compounds like methane and carbon monoxide that seem more conducive to the formation of life than planetary and lunar environments in the inner solar system

2,200	2,100	2,000	1,900	1,800
Widespread occurrence of prokaryotic plankton (bacterioplankton) in world's oceans.	Increasing UV (ultra-violet ray)- absorbing ozone shield (O_3 derived from O_2 accumulating in atmosphere.) Oldest abundant fossil bacteria: *Gunflintia, Huronospora, Leptoteichus golubicii,* etc.	Free O_2 abundant in atmosphere, indicating dominance of aerobic organisms. Mitochondria, ancestors to most eukaryotes, acquired by symbiosis as purple eubacteria. Gunflint Iron Formation (present-day Ontario, Canada) and equivalent fossil biotas in present-day China, Australia, and California containing complex filamentous microfossils and remains of structured communities.	First appearance of *Grypania,* identified as the earliest Protoctista (maybe reinterpretable as discarded cyanobacterial sheaths).	Replacement of banded iron formations by red beds (oxidized iron sediments), indicating worldwide transition to an atmosphere rich in oxygen.

"tholins," organic goop formed in the laboratory of astronomer Carl Sagan at Cornell University under conditions simulating those thought to exist on the similarly colored clouds of Jupiter.

Meanwhile, British chemist Graham Cairns-Smith has proposed that clays would have shielded fragile cell precursors from the solar rays which, though capable of assembling organic compounds, could also irradiate them into oblivion.[10] Crystalline clays, others have suggested, could have aggregated on nonliving bubbles produced by wind, rain, volcanoes, and waves. Even today, attracting particles along their surfaces while undergoing changes in temperature and pressure, bubbles serve as a meeting place where ambient carbon, nitrogen, hydrogen, and other elements form more complex compounds. When bubbles burst they leave chemical residues in their wake.

Whatever the precise route taken in the origin of life, Freeman Dyson proposes that it probably came by way of a kind of molecular "symbiosis" (although the word is not quite right, since neither partner was itself alive) between RNA—a "supermolecule" probably crucial, as we shall see, to life's origin—and more haphazardly growing "protein creatures."[11] Despite much conjecture and intriguing research, it must be remembered that no life has yet been synthesized in the laboratory. The gap between chemical evolution (the appearance of carbon compounds in "environmental" mixtures) and true cells (self-bounded, self-maintaining, and ultimately reproducing matter) remains unbridged. Nonetheless, the laboratory-explored shenanigans of RNA—as we shall see in a moment—are shrinking this bridge daily.

"Stumbling Forward"

Humanity's understanding today of the origins of life is probably no better than was our understanding fifty thousand years ago of the origins of fire. We can maintain and play with it, but we can't yet start it. The assumption that the origin of life may be repeated by investigators in the laboratory is a shocking example of the audacity of scientists—and yet it may prove correct.

Scientific investigation does reveal gradations between certain chemical systems and the animated material all of us recognize as life. Schrödinger's crystal analogy has given way to an idea of life as a chemical system requiring material and energy to persist far from thermodynamic equilibrium, i.e., a dissipative system. Dissipative systems that are not alive may nevertheless act in ways that are eerily lifelike. One such dissipative system develops in the Belousov-Zhabotinsky reaction. It involves the oxidation of malonic acid by bromate in a sulfuric acid solution containing cerium, iron, or manganese atoms. [PLATES 17 AND 18] Under certain conditions, concentric and rotating spiral waves will occur in an aesthetically captivating chemical reaction that may last for hours.

The regularity and duration of such reactions have led some scientists to compare them with life. Using energy from outside to increase their internal order, these chemical systems, some of which are brightly colored, "live" for a while beyond the limit of equilibrium chemistry. Erich Jantsch, an Austrian-American astrophysicist and philosopher, explains that,

> Whereas free energy and new reaction participants are imported, entropy and reaction

Appearance of second kingdom: PROTOCTISTA.

Earliest eukaryotes documented in fossil record as acritarchs, indicating cell evolution by symbiosis.

Origin of speciation inferred from molecular data on protoctists (primarily anaerobic mastigote protists).

Diversification of aerobic life.

Appearance of planktonic and benthic organisms possibly correlated to symbiotic acquisition of air-breathing mitochondria.

Protoctist evolution: origins of mitosis, meiotic sex, gender, and programmed death of individuals in eukaryotic microorganisms and their descendants.

Appearance of terrestrial cyanobacterial life (desert crust and soil microbial communities).

Diversification of seaweeds (algae, which are photosynthetic protoctists) of unknown taxa possibly correlated to symbiotic acquisition of photosynthetic plastids.

17 (above)
Proteus mirabilis, a bacterium. Phylum: Omnibacteria. Kingdom: Monera. Nested patterns of concentric terraces produced by *Proteus mirabilis.* The bacteria form these patterns by repeated cycles of growth and migration over the petri plate's agar surface. The living geometry of this photograph is reminiscent of the non-living dissipative structures of plate 18.

18 (below)
Autocatalytic chemical reactions of the same sort but using different substances as recursive structures are thought to give rise to life. This particular "chemical clock" is a dissipative structure in a Belousov-Zhabotinsky reaction. The increase in complexity over time is reminiscent of life. By reproducing, however, life has increased its complexity not for several minutes but for several thousand million years.

end products are exported—we find here the *metabolism* of a system in its simplest manifestation. With the help of this energy and matter exchange with the environment, the system maintains its inner non-equilibrium, and the non-equilibrium, in turn, maintains the exchange processes. One may think of the image of a person who stumbles, loses his equilibrium and can only avoid falling on his nose by continuing to stumble forward. A dissipative structure continuously renews itself and maintains a particular dynamic régime, a globally stable space-time structure. It seems to be interested solely in its own integrity and self-renewal.[12]

Dissipative structures, chemical systems that use streams of energy to increase their internal order, are, however, rare and short-lived. But if the increased internal order is that of life, then, given access of the system to a source of energy and the right kind of matter (nutrients), it maintains indefinitely. This is autopoiesis. Autopoiesis is what happens when a self-bounded chemical system—based not on small molecules of sulfuric and malonic acids but on long-molecule nucleic acids and proteins—reaches a critical point and never stops metabolizing.

The cell is the smallest autopoietic structure known today, the minimal unit capable of incessant self-organizing metabolism. The origin of the tiniest bacterial cell, the first autopoietic system, is obscure. Yet most agree that complex carbon compounds, exposed in some way to unceasing energy and environmental transformation, became oily droplets that eventually became membrane-bounded cells.

| Continued diversification and widespread appearance of monera (iron bacteria, cyanobacteria, and many unknown Protoctista—sexual cysts, algae, microscopic and even large fossils.) | Global rifting event. | Increase in diversity of algae (photosynthetic eukaryotes) and other protists. | Oldest "giant" acanthomorph acritarchs, probably algae. | Worldwide proliferation of unidentified large "quilted" organisms, fossilized in sandstone, probably colonial sand-dwelling members of diverse protoctist kingdom, the Ediacaran biota. |

Metabolism, the chemical measure, the specific earthly manifestation of autopoiesis, has been a property of life since it began. The first cells metabolized: they used energy (from light or from a small range of chemicals—never from heat or mechanical movement) and material (water and salts, carbon, nitrogen, and sulfur compounds) from outside to make, maintain, and remake themselves. Autopoiesis, the chemical basis for the impatience of living beings, is never optional. Absolutely required at all times for any life form in a watery milieu, once autopoiesis appeared in the tiniest bacterial ancestor, it was never completely lost.

You embody the processes of the early Earth in your living cells. The failure of the autopoietic system of cell maintenance is death. If autopoiesis of a cell ceases, the cell dies. A many-celled organism capable of replacing its cells survives as the autopoietic behavior of the larger organic being prevails. If too many component cells die, metabolism of the larger entity halts and death follows. Any cell or organism that continues to self-maintain will grow, and the imperative to reproduce will follow. Although not obvious to the naked eye, cell metabolism never stops. Chemical trans-formations such as nutrient uptake and energy conversion, and the fabrication of DNA, RNA, and proteins, occur continuously in all cells and all beings made of cells.

Life seems to have originated in whatever were the primordial ancestors of modern bacteria. Chemical systems that became biological systems, these first beings would have metabolized and incorporated energy, nutrients, water, and salts into their developing selves. The first cells formed. As in Jantsch's analogy of the person stumbling forward to avoid falling on his face, so membrane-bounded cells replicating RNA and producing other molecules stumbled onto DNA-based RNA and protein synthesis; that is, reproduction became a means to retain self-maintenance, to postpone return to thermodynamic equilibrium.

Bacteria reproduce in the time it takes to read this chapter. Elephant and whale reproduction can require a decade. But whatever the rate, reproduc-tion requires DNA replication in cells. It requires RNA, protein, and membrane synthesis and the intrinsic locomotion of growth. Reproduction of larger beings—protoctists, fungi, animals, and plants—also involves growth and division of their component cells. Autopoietic multicellular beings are composed of cells which themselves are autopoi-etic. Animal and plant reproduction is a permuta-tion of cell autopoiesis, just as cell autopoiesis is a permutation of nucleic acid and protein metabo-lism. Our instinctive desire to live is directly related to the autopoietic imperative to survive, itself related to the "yearning" of heat to dissipate.

Metabolic Windows

Because cells retain their organization in spite of—or because of—the helter skelter around them, they provide science with a window onto the past. It is a rather magical fact that, within the autopoietic, thermodynamic view, our bodies today should have virtually the same chemistry as that prevailing on Earth's surface three thousand million years ago. Remember that when life became autopoietic it postponed indefinitely the moment of total heat equalization and loss of order. Using the energy of food and sunlight, thermodynamic equilibrium has been thwarted.

PHANEROZOIC EON (Beginning of "Human-centered" Timeline.)

Series of worldwide "Vendian" ice ages followed by diverse new planktonic and benthic communities, probably of protoctists.

Appearance of third kingdom: ANIMALS.

Inferred origins of egg, sperm, embryo, and blastula.

Appearance in fossil record of soft-bodied animals (sponges, coelenterates, arthropods, and others).

Paleozoic Era.

Cambrian, Ordovician, Silurian, and other time-rock periods (see page 54).

"Age of Marine Animals".

Trilobites and other hard-shelled animals prevalent.

Appearance of fourth and fifth kingdoms: PLANTS and FUNGI.

Mesozoic Era.

Triassic, Jurassic, and Cretaceous Periods.

"Age of Reptiles".

Death is illusory in quite a real sense. As sheer persistence of biochemistry, "we" have never died during the passage of three thousand million years. Mountains and seas and even supercontinents have come and gone, but we have persisted.

We have, of course, had to "up the stakes" at various junctures to stay alive. This continuous "upping of the stakes," which, on the personal level, links desire to death, is on the species level described as evolution. Beings always require food and energy to stay the same, and often they have to evolve, to change into new forms, simply to self-maintain. The feline lineage, the flowering plant lineage, the nautiloid-squid and the rest of the cephalopod lineage have changed and persisted through the sexual reproduction and death of their members.

Evolution, no less than the nucleic acid replication of autopoiesis and reproduction, is a "stumbling forward" to stave off the threat of thermodynamic dissolution. Most atoms in our bodies are hydrogen—the element which as a gas was, according to astronomic models, blasted beyond the confines of the inner solar system when the sun turned on. Nonetheless, these atoms which should be long gone have defied time and space by becoming bound up with (as) life. Today hydrogen-rich gases such as ammonia exist not only in the atmospheres of the giant outer planets but in the inner solar system where life has preserved them in its self-similar structure ever since it began maintaining and reproducing.

Indeed, the original dissipative chemistry, the protein and nucleic acid chemical clocks that arose prior to life may even have been preserved. One of the most beautiful aspects of living things is that they bear within their very form the presence of the past. We resemble our parents and other people who lived ten thousand years ago. This preservation of the past in the present is fortunate for scientists. Each body is the charitable gift of a biochemical museum, and each bacterial cell an unplanned time capsule.

Far from lost in what Shakespeare called the "dim backward and abysm" of time, life's origins are an open secret awaiting deciphering by sufficiently talented chemists. If life is an autopoietic, far-from-equilibrium phenomenon, living cells should still contain significant fragments of preliving systems. Vestiges of life's origin may still exist, a stuttering genesis for scientists patient enough to listen. Life may even contain the original dissipative structures and chemical fossils in the form of metabolic pathways. Ultimately far more valuable than micro-fossils, or the modern alchemical experiments of energizing chemicals in laboratory glassware, are organisms: easy to overlook in their obviousness—uncannily present—they are metabolic windows onto life's origin.

The RNA Supermolecule

The minimal free-living autopoietic entity today is probably a tiny, spherical, oxygen-shunning bacterium which requires energy and food to keep going its 3,000 genes and proteins. Or maybe it is a kind of mycoplasm, a being so small that until recently it was known only as a cousin to a growing speck that caused disease symptoms in brains of sheep. Even in these, the bonded atoms of carbon, hydrogen, nitrogen, and oxygen interact recursively in a metabolic system.

The genes that are DNA require active RNA to work. DNA and RNA together make the proteins that

200   100   65   4

Present

Cenozoic Era.
"Age of Mammals".

Appearance of
human ancestors.

19 (next page)
The backbone of the feather-like structure is composed of protein-coated DNA actively being transcribed into messenger RNA molecules, which form the outward growing strands in this photograph taken with an electron microscope. This activity of the genes—DNA molecules using messenger RNA to make the thousands of proteins of all cells from the pseudopods of amebas to the feathers of parakeets—is the *sine qua non* of life as we know it on this planet. DNA copies itself to make genes but it is also transcribed into messenger RNA that puts together the rest of the organism.

form cell structures, and they also make the very enzymes that slice and splice the genes. The so-called genetic code actually refers to the correspondence between the linear order of DNA's components and that of the amino acids in a vast array of different proteins.

With the help of RNA, the nucleotides of DNA line up amino acids of protein. Our blood, internal organs, finger nails, skin, and hair are all made of proteins. [PLATE 19] The reason that nutritionists advise us to eat the "eight essential amino acids" is because the human body cannot renew itself without taking in these protein components from food and drink. The human body cannot synthesize these particular amino acids at all—even their simpler components.

In contrast to our human need to mine the environment for essential amino acids, no living being on Earth needs to stalk the environment in quest of the deoxyribose sugar essential for its DNA. Rather, deoxyribose is synthesized in cells by diverting an oxygen atom from ribose. It is ribose, the five-carbon sugar of RNA, that is often taken up from the outside as food. That all cells, given ribose sugar, can make deoxyribose from it suggests that ribose came first. RNA, with ribose, evolved before DNA. DNA sugar metabolism evolved by subtraction of oxygen from RNA sugars. The earliest cells may have been RNA beings which only later evolved DNA systems. Comparing RNA and DNA metabolism is an example of peering into cellular windows for clues to life's most ancient origins.

Other evidence questions the pretensions of DNA, the "master molecule," to life's biochemical throne. RNA, more versatile than DNA, is a better choice for the replicative tool of life's earliest autopoietic

system. Whereas double-stranded DNA uses deoxyribose sugar for its chain, single-stranded RNA uses ribose, the source material of deoxyribose. Unlike DNA, which must use RNA to code for proteins, RNA by itself can direct both its own replication and the making of proteins. In ancient times, RNA probably did all that DNA does today inside of cells, and more. In all cells, when the two helically coiled strands of DNA open to expose a section of the nucleotide sequence, that portion of the DNA is "copied" onto messenger RNA. Taking its message to two other kinds of RNA (transfer RNA and ribosomal RNA—named for the ribosomes, the "factories" in the cell where proteins are made), messenger RNA's information is "translated" into the amino acid units that assemble into working proteins. RNA can make proteins, in principle, without any DNA.

Following the lead of Sol Spiegelman at the University of Illinois in the late 1960s, German Nobel physicist Manfred Eigen (together with coworkers at the Göttingen Institute) found a way to induce test tube RNA molecules to replicate by themselves. Eigen showed that nucleotide units of RNA lined up and formed functional RNA. Most impressively, some of the test tube RNA even mutated into a different RNA that replicated more quickly than the original. The Eigen experiment did not, of course, reveal the spontaneous generation of life; RNA molecules by themselves are not cells. RNA in test tubes would have remained completely lifeless had scientists not extracted proteins from live cells and added them to test tubes containing RNA.

Eigen's RNA molecules are much like viruses. Certainly not alive, they show a power on the border

of life. As computer viruses require human-run computers to spread, so naturally occurring viruses—not full, autopoietic beings but genes coated with protein—require living cells. Replicating RNA viruses may be just as dangerous and capable of replication as are DNA viruses.

Donald Mills of Columbia University also made test tube RNA viruses; these RNA viruses used a bacterial enzyme to replicate themselves inside a bacterium Mills had conveniently provided for them. In the early 1980s Thomas Cech of the University of Colorado and Sidney Altman of Yale University found that certain kinds of RNA are capable of self-splicing. In recognition that RNA acts like its own active protein by cutting and rearranging itself the way enzyme proteins do, these test-tube reactants are called "ribozymes." Ribozymes, cell-free mixtures of the right pieces of RNA, proteins, and their components, furthermore change with time in a kind of test-tube evolution. Gerald Joyce at the University of California, San Diego, may be on to the hottest biochemical project yet: with his colleague Jack W. Szostak he has found ribozymes for which the associated enzymes actually accelerate the rate of RNA replication: veritable molecular evolution in a test-tube.[13]

RNA is thus a prime candidate for early life's supermolecule. By working on itself as it grew, RNA may have generated a mélange of expanding possibilities. Able to replicate and mutate, acting as both enzyme and gene, RNA performs operations that form more of itself.

On the Archean Earth, we can now envision that it was bounded RNA that developed replication and information systems, as, sequestered inside oil droplets, it began the trek toward autopoiesis. The DNA world of all life today may thus have evolved inside RNA cells of the "RNA world," a term coined in 1986 by Harvard University biologist and Nobel laureate Walter Gilbert.[14]

Cells First

In the normal waking state, human bodies burn sugars aerobically, using oxygen atoms drawn from the air. But in strenuous activity the body reverts to a distinct metabolism; muscles ferment sugars in the same anaerobic way invented by early bacteria. When stressed, our bodies thus "remember" the times before the atmosphere became suffused with oxygen. Such physiological flashbacks re-present past environmental conditions and the bodies that evolved to live in them. In a very real sense, all beings today retain traces of Earth's earliest biosphere.

Neither DNA nor RNA alone is enough to form life. American biophysicist Harold Morowitz, using the concept that "metabolism recapitulates biogenesis" (biogenesis referring to the origin of life), suggests that nearly all aspects and forms of metabolism and the synthesis of proteins and nucleic acids evolved only after membranes enclosed the precursors of cells. Whether proteins or nucleic acids "came first" in the sequence of life's origins, membranes surely arose even earlier. Life is thus truly a cellular phenomenon.

Morowitz emphasizes that, in the watery environment in which life arose, a nonaqueous barrier was necessary to separate the cell from its surroundings. "To be an entity, distinguished from the environment, requires a barrier to free diffusion. The necessity of thermodynamically isolating a subsystem is an irreducible condition of life....It is the closure of an amphiphilic bilayer membrane into a vesicle that represents discrete transition from nonlife to life."[15] Considered materially, as a system of matter and energy, life is recognizable by its partial separation from the environment by way of a membrane.

Over evolutionary time, individuality, always based on the membrane-bounded unit of the cell, arises at ever greater levels of integration. Life not only evolves, it epitomizes. A cell membrane is common and necessarily intact in all autopoietic beings. A membrane is a precondition for cell metabolism. The first autopoietic system, which may have lacked both DNA and RNA, was almost certainly a cell. It may have become an RNA-filled cell, a lipid capsule whose autopoietic persistence, enhanced by the addition of DNA, evolved into our earliest bacterial ancestor.

At the beginning of this chapter we wrote that life had been abandoned by the world, yet the world had gone nowhere. We hope to have shown that this was more than poetic metaphor: life was in the midst of the world and matter but separated from it by a translucent, semipermeable membrane.

So, what is life?

Life is the representation, the "presencing" of past chemistries, a past environment of the early Earth that, because of life, remains on the modern Earth. It is the watery, membrane-bound encapsulation of spacetime. Death is part of life because even dying matter, once it reproduces, rescues complex chemical systems and budding dissipative structures from thermodynamic equilibrium. Life is a nexus of increasing sensitivity and complexity in a universe of parent matter that seems stupid and unfeeling in comparison. Life must maintain itself against the universal tendency of heat to dissipate with time. This thermodynamic view explains, in a way, the determination, the purposefulness of life — for billions of years it has been stuck in a pattern which, even if it wanted to, it can't get out of, of upping the stakes as it goes. For life itself is — are — these patterns of chemical conservation in a universe tending toward heat loss and disintegration. Preserving the past, making a difference between past and present, life binds time, expanding complexity and creating new problems for itself. ■

4 Masters of the Biosphere

Had [bacteria] been discovered on Mars, their description would have been much more dramatic and the bizarre quality of their natural history, which often seems like science fiction, would not have been missed.

—**Sorin Sonea** and **Maurice Panisset**

Perhaps some of my readers will respond with a smile to my doctrine of living contagions.

—**Agostino Bassi**

Fear of a Bacterial Planet

Microorganisms were a curiosity, a kind of natural history sideshow, before it was realized that some cause disease. Antoni van Leeuwenhoek, inventor of an early version of the microscope in the 1670s, described these beings as "animalcules"—tiny animals. He was struck by their rapid movement, odd shapes, and sheer quantity. In 1831 the half-blind Italian law student Agostino Bassi (1773-1856) proved the existence of infection by spreading muscardine (silkworm disease) from one fungus-infected worm to another. Nonetheless, a generation after Bassi showed that disease did not arise spontaneously, even Pasteur thought of bacteria only as agents of decay.

A turning point occurred when Robert Koch (1843-1910) found bacteria in blood of cows stricken by anthrax. These little rods ("bacilli") grew from hardy bacterial spores. Feeding them blood serum, Koch, a German medical officer, learned to grow the bacteria in a liquid broth. He developed a stain for them, photographing the culprits by mounting a camera on a microscope. Yet the now-common notion that bacteria caused infectious disease was slow to be accepted. English nurse and philanthropist Florence Nightingale (1820-1910) denied the existence of germs to her death.

When the germ theory of contagion finally caught on, it did so with a vengeance. Different types of bacteria were implicated in anthrax, gonorrhea, typhoid, and leprosy. Microbes, once amusing little anomalies, became demonized. Pasteur, like Howard Hughes after him, had a phobia about dirt and germs. He avoided handshakes. Wiping down his crockery, he meticulously sought evidence of wood, wool, and other detritus in his food. No longer an amusing scandal for parlor conversation, microbes became a virulent "other" to be destroyed. The metaphor of tenacious, infectious bacteria was used in Nazi rhetoric of genocide.

Today the low regard for bacteria as Lilliputian "agents of disease" still obscures their enormous importance to the well-being of all the rest of life.

Until the 1950s the oldest unambiguous fossils were 550-million-year-old trilobites and other extinct marine animals. The oldest rocks on Earth, by contrast, dated to nearly four thousand million years. Fossils of microorganisms have now been confirmed in the oldest sedimentary rocks, suggesting that life took root soon after the Earth-Moon system formed.

In 1977 paleobiologists Elso Barghoorn and Andrew Knoll of Harvard University found some two hundred fossil bacteria (some even in stages of cell division) in sedimentary rock 3,400 million years old. Because Barghoorn had earlier discovered microbial life in the Gunflint Iron Formation of western Ontario and states bordering Lake Superior, he was well prepared to detect the fossil remains of bacteria in these vastly older African and Australian rocks. In 1990 geologist Maud Walsh went to the Barberton Mountain Land in southern Africa and collected ancient black rocks called cherts, silica-rich rocks petrified from mud flats and volcanic pools. Back in the Baton Rouge, Louisiana, laboratory, the cherts were sliced and polished into thin sections for examination under a microscope. Walsh saw more than just bacteria; she saw bacteria trapped in sandy layers of their own making, testifying to whole microbial mat communities that had flourished 3,500 million years ago.[1]

Modern bacteria may offer even more revealing clues about earliest life. American molecular biologist Carl Woese has found that three types of very tough bacteria are distinguished from all others by their ribosomal RNA: salt-loving "halophiles," heat-loving "thermophiles" of hot springs, and methane-producing "methanogens." These extreme-condition dwellers have ribosomal DNA that makes them more similar to one another than to all other bacteria. Woese calls the hardy beings "archaebacteria"; he suggests they are direct descendants of earliest life on Earth.

The observation that archaebacteria inhabit oxygen-free environments — such as ocean bottom, stomachs of cows, oxygen-depleted sewer water, and the hot acidic springs of Yellowstone National Park — agrees with the contemporary picture of a hot Archean Earth with, at most, traces of oxygen in the atmosphere. Oxygen was only released into the atmosphere once blue-green bacteria evolved a way to use energy from sunlight to break apart water molecules (H_2O) to grab the precious hydrogen. Combining the hydrogen with carbon atoms drawn from then-abundant carbon dioxide, blue-green bacteria were able to manufacture DNA, proteins, sugars, and all their other cell components. These light-needy bacteria quickly expanded to sunny waters everywhere on the Archean Earth. In so doing, they released vast amounts of molecular oxygen left over from their hydrogen mining of water.

Earth's atmosphere thus became an extension of the metabolism of evolving bacteria. Only through the workings of the most innovating bacteria of all time did the originally anoxic Earth gain an oxygen-rich atmosphere. The planet had first been populated by methane-makers, sulfur-lovers, and other anaerobes — beings that neither produced nor used oxygen gas in their metabolism.

Life Is Bacteria

One legitimate answer to the question "What is life?" is "bacteria." Any organism, if not itself a live bacterium, is then a descendant — one way or another — of a bacterium or, more likely, mergers of several kinds of bacteria. Bacteria initially populated the planet and have never relinquished their hold.

Bacteria may be Earth's tiniest life forms, but they took giant steps in evolution. Bacteria even invented

20
Myxococcus, multicellular bacterium. Phylum: Myxobacteria. Kingdom: Monera. These bacteria aggregate into multicellular "trees" when nutrients or water are depleted. Contrary to common opinion, "multicellularity" is not a feature solely of "higher" organisms but is already an entrenched feature of prokaryotes.

multicellularity. Most bacteria in nature, in spite of popular belief, are multicellular. [PLATE 20] In these multicellular bacterial beings, each unit cell itself is a bacterial cell.

Some lineages of bacteria went on to evolve into many different kinds of beings, including ourselves. Inside the cells of all of us right now are former bacteria, using oxygen to generate energy. These are the mitochondria. The photosynthetic blue-green beings and their descendants (plants) remove carbon dioxide from the atmosphere, using carbon for their bodies and eliminating as waste the oxygen of fresh air—only a small portion of which is used by the mitochondria that cohabit with what were photosynthetic bacteria in all plant cells.

Our neighbor planets, Mars and Venus, both have atmospheres more than 90 percent carbon dioxide. Yet Earth's air is a reactive mixture, with its huge portion of oxygen and less than 0.1 percent carbon dioxide. It was bacteria that removed the carbon dioxide and produced the oxygen. Bacteria, truly, have made the planetary environment what it is today. All larger beings contain mitochondria within their cells, living descendants of bacteria that lived on Earth before oxygen had accumulated in the air. Life on Earth is a holarchy, a nested fractal network of interdependent beings.

The fear of bacteria is, in a way, a fear of life, of ourselves at an earlier stage of evolution. It is not surprising that microbes now find us so attractive. Because the carbon-hydrogen compounds of all organisms are already in an ordered state, the human body (like that of every other living being) is a desirable food source for these tiny life forms.

1

Bacteria seek us as a source of autopoietic maintenance in their ancient struggle against thermodynamic equilibrium.

Perhaps we should take solace in the fact that the matter of our bodies returns upon death not to an inert state of matter but to the bacterial order undergirding the biosphere. "Don't you see," wrote Giordano Bruno, "that which was seed will get green herb, and herb will turn into ear and ear into bread. Bread will turn into nutrient liquid, which produces blood, from blood semen, embryo, men, corpse, Earth, rock, and mineral and thus matter will change its form ever and ever and is capable of taking any natural form."[2]

The wish to maintain one's youth, one's most attractive form, and ultimately one's very life is thwarted at the level of the animal body. But our individual defeat is a victory for the bacteria, which return the hydrogen-carbon compounds of our bodies to a living environment. Closer to life's original structures, bacteria do not live as we live, toward death. Banning an unfortunate accident, a mutation, or a gene-trading encounter with another bacterium, a single bacterial cell can "survive" essentially forever in its original form, as generation after generation of bacterial copies of itself are made by cell division.

We pluricellular creatures are each a disequilibrium structure of cells, as a bacterium is a disequilibrium structure of matter. Humanity as a species, even the entire kingdom of animals, has a far more fragile existence than theirs—just as their existence is more tenuous than that of nonliving matter.

The Metabolically Gifted

Bacteria can swim like animals, photosynthesize like plants, and cause decay like fungi. One or another of these microbial geniuses can detect light, produce alcohol, waft hydrogen and fix nitrogen gas, ferment sugar to vinegar, convert sulfate ions or sulfur globules in salt water to hydrogen sulfide gas. They do all this and much more not because they are "pathogens" or in service to clean our human environment but because their survival imperative led to their inventing every major kind of metabolic transformation on the planet.

The smallest of them have a diameter only a thousand times greater than a hydrogen atom. If there were such a thing as angels that could dance on the head of a pin, bacteria would be they. Bacteria have already mastered nanotechnology; already miniaturized, they have control of specific molecules about which human engineers dream. Far more complex than any computer or robot, the common bacterium perceives and swims toward its food. Choosing and approaching its destinations, bacteria propel themselves by flagella, corkscrew-shaped spinning protein filaments attached to living motors in the membranes of their cells. Complete with rings, tiny bearings, and rotors, they are called "proton motors" and spin at about 15,000 rpm. These proton motors move bacteria in the same way that "electric fan" outboard motors propel boats.

Rapidly reproducing, bacteria properly supplied with food and water double their cells in a half hour or faster. They have been and probably always will be the most important players in maintaining the biosphere. A single photosynthetic blue-green bacterium growing and dividing under ideal conditions could, in theory, produce all the oxygen now in the atmosphere in just a few weeks.

All other life forms depend on the workings of uncountable living, dying, and metabolizing bacteria. Our relations to the bacteria all around us factor into our health and well-being and that of our soil, food, and pets. As naked and simple as bacterial cells seem, they are continuously busy at the cellular and the planetary level. Varieties of bacteria exist that photosynthesize without ever releasing oxygen, yet they use carbon dioxide and hydrogen to produce all their cell parts. Others incorporate carbon dioxide and hydrogen from the air into body protein, converting their waste into methane gas. Still others turn sulfate to sulfide, or incorporate inert nitrogen into their bodies. Only citizens of the bacterial kingdom are so metabolically gifted. When an animal (like the termite who produces methane) or a plant (like the starved bean who begins to supply itself with nitrogen from its roots) is discovered with such metabolic skills it is because they have co-opted the bacterial bodies to their expertise. Such borrowing also applies to biotechnology performed by humans in white lab coats. We humans do not "invent" patentable microbes through genetic recombination; rather, we have learned to exploit and manipulate bacteria's ancient propensity to trade genes.

The Gene Traders

Bacteria trade genes more frantically than a pit full of commodity traders on the floor of the Chicago Mercantile Exchange. The trading by bacteria of genetic information provides the basis for understanding new concepts of evolution.

Evolution is no linear family tree but change in the single, multidimensional being that has grown now to cover the entire surface of Earth. This planet-sized being, sensitive from the beginning, has become more expansive and self-reflexive as, for the past three thousand million years, it has evolved away from thermodynamic equilibrium. Imagine that in a coffee house you brush up against a guy with green hair. In so doing, you acquire that part of his genetic code, along with perhaps a few more novel items. Not only can you now transmit the gene for green hair to your children, but you yourself leave the coffee shop with green hair. Bacteria indulge in this sort of casual quick-gene acquisition all the time. Bathing, they release their genes into the surrounding liquid. If the standard definition of

species, a group of organisms that interbreed only among themselves, is applied to bacteria, then all bacteria belong worldwide to a single species. The Archean Earth was a promiscuous place of prodigious growth and rapid gene transfer that led, by and by, to the genetic restrictions of the Protero-zoic protists, the larger composite beings presented in our chapter 5.

Unlike all familiar sexually reproducing species, whose members have cells with nuclei in them that package their DNA, the DNA of bacteria is loose inside their bodies. Bacterial cells entirely lack nuclei; for this reason bacteria are prokaryotes composed of prokaryotic cells. "Prokaryote" literally means "before nuclei." Free of nuclei, and unfettered even by the red-staining, protein-coated chromosomes of all other forms of life, bacteria never reproduce by mitosis. Mitosis, the "dance of the chromosomes," is the kind of cell division by which the cells of plants, fungi, and animals always divide. This dance evolved in protists of the Proterozoic, the eon that followed the Archean. By contrast, a parent bacterium elongates its DNA dragged by growing membrane to which it is attached until the full-grown cell splits to form two offspring identical to it. Some bacteria reproduce by "buds," protrusions on the single parent that yield smaller offspring, all of which contain the parent's same genes.

Members of familiar species of plants and animals reproduce "vertically," as mother and father each donate an equal number of genes (on chromosomes) to form new offspring. Bacteria are under no such constraint. Rather, bacteria trade genes "horizontally," acquiring new genes from peers in their own generation.

Bacterial cells often have spare strands of DNA, that is, extra sets of genes. These genes may be traded in naked pieces called plasmids or as protein-coated pieces called viruses. In some bacteria a cell bridge forms between the one donating its genes and the one receiving them. [PLATE 21] This process of growing a cell bridge through which genes are sent, called conjugation, is distinct from mammalian sex. No bacterial cells fuse nor do "parents" make equal contributions to an offspring. Rather, one bacterium, the "donor," passes its genes in one direction to the "recipient," which does not reciprocate the favor. Still, this conjugation meets the minimal requirements of a biological sex act since the transfer of genes produces a new bacterium, a "genetic recombinant" being with genes from more than a single parent.

Bridge-forming bacterial conjugation is limited. Many types of bacteria that cannot conjugate indulge in viral or plasmid sex. Those that practice this more common form of sex require a difference between the bacterial "genders": a donor needs a recipient. Whether any given bacterial cell is donor or recipient is determined by a single "sex" gene. The sex gene may itself be transferred in the conjugation process. If this occurs, a "male" (donor) bacterium can become a "female" (recipient). "She" becomes a male donor like "himself." Any number from a very few to a very many genes may be transferred at a time, conferring on the recipient not just an ability to make cell bridges but other useful traits, such as an ability to manufacture vitamins or to resist a particular antibiotic.

When exposed to ultraviolet radiation, healthy bacteria explode with tiny viruses called prophages. Such viruses spread genes to surviving recipients. Because on the early Earth atmospheric ozone was not around to intercept the sun's ultraviolet rays, genetic exchange may have been even more prevalent than it is today. The early UV-bombarded Earth may have been the scene of a multimillion-year orgy of gene trading bacterial sex.

Bacterial recombination is a natural form of the "genetic" recombination exploited by biotechnologists. Manipulating a preexisting bacterial penchant, technicians force the colon

21
Three-way genetic exchange among bacteria. Unlike all other forms of life on Earth, bacteria transmit genetic information relatively freely such that taxonomically different "species" can trade genes. Bacterial sex, important to the evolution of cells with nuclei (eukaryotes) was probably rampant before bacteria themselves produced sufficient oxygen gas to create an ozone layer. The male in the upper left of the electron micrograph is sending genes through two tubes covered by bacteriophage viruses.

Chromatium vinosum, purple sulfur bacteria. Phylum: Anaerobic Phototrophic Bacteria. Kingdom: Monera. These photosynthetic microbeings photosynthesized in the light long before plants evolved. The purple inclusions are the thylakoid membranes of photosynthetic pigments and enzymes, the spheres are sulfer. These beings, which only tolerate oxygen in the dark, testify to the fact that photosynthesis is an anaerobic process that evolved long before oxygen was present in the air.

in retinal rods of marine fish and is used in our own vision, especially when light is dim.

The time-travelling eye, moving forward in time somewhat beyond the origins of the green sulfurs and the red halophiles, would eventually spot a new photosynthetic form: purple sulfur bacteria. [PLATE 24] Their brilliant innovation was not so much a differently pigmented approach to photosynthesize but an ability to tolerate oxygen. Their tolerance was not, however, complete. Purple sulfur bacteria living today can tolerate oxygen only in the temporary darkness of nightfall or other cover. Yet photosynthetic beings, given their autopoietic imperative, must dwell in sunlight; none can live for long in the dark.

Oxygen Excitement

The purple sulfur bacteria of the Archean that could tolerate oxygen had an advantage. Oxygen was now beginning to creep into the environment. A new form of water-using bacteria, the cyanobacteria, had put it there.

Sometimes still called plants or "blue-green algae," the cyanos are neither plants nor algae. Cyanobacteria wreaked havoc with the planetary environment—more so than has any life form before or since. Life had always existed in a medium rich in hydrogen: H_2O, water. Nonetheless, life's supply of hydrogen to make the organic compounds of its bacterial bodies had heretofore come from sugars such as glucose ($C_6H_{12}O_6$), or from hydrogen and hydrogen sulfide in the air. The cyanos evolved when photosynthetic bacteria, employing a unique green chlorophyll system, mutated from purple predecessors to get their hydrogen atoms from water. Splitting apart hydrogen dioxide (water) into its constituent atoms, the blue-green bacteria assembled hydrogen into themselves.

Far more abundant than stinking hydrogen sulfide, clear water abounded. Wherever they could access water and sunlight, the blue-green bacteria grew. Today in light and water, these oxygen-making photosynthesizers, still capitalizing on their ancient metabolic innovation, continue to thrive. More than ten thousand kinds have been cataloged. They are found virtually everywhere—on damp, dimly lit walls at the mouths of caves, following the slow leaks of refrigerators, on boat decks, boulders, cliffs, drain pipes, toilet tanks, and shower curtains. They exist in the Red Sea, in boiling springs, nuclear reactor cooling tanks, the Sinai Desert, across the Siberian tundra, and under the Antarctic ice. Cyanos have been cast by some scientists as the beings most likely to reproduce were they to be strewn about the red surface and polar dry ice caps of Mars.

The rampant growth of cyanobacteria on Earth was not a local phenomenon. Wherever blue-green cyanobacteria grew they incorporated the "H" from H_2O into their tiny bodies and released into the air the "O" as O_2, oxygen gas. Highly destructive to all cells when it causes miniature biological explosions, oxygen gas was fatal to most forms of early life. Even today it is toxic in high concentrations. Oxygen combines dangerously with enzymes, proteins, nucleic acids, vitamins, and lipids. And oxygen produces "free radicals"—reactive, short-lived chemicals that interfere with metabolic systems. Nutritionists have implicated free radicals in the human aging process, recommending anti-oxidants such as vitamin E.

In the Archean eon oxygen reacted, sometimes violently, with atmospheric gases such as hydrogen, ammonia, methane, and hydrogen sulfide. Origins-of-life theorists agree that life on Earth has little to no chance of re-evolving here because free oxygen would oxidize hydrogen-rich chemicals crucial to starting any life. The outer solar system, however, is a different story. Free oxygen is not a major constituent of the atmospheres of planets like Jupiter or moons like Titan. Indeed, if life were to evolve again it would be far more likely to do so in the gaseous hydrogen-carbon environments of the outer solar system, where free oxygen is not around to interrupt early life's oxygen-intolerant chemical systems.

Releasing oxygen, the blue-green bacteria enveloped the world with oxygen atoms. Oxygen accelerated change as it sped up chemical reactions. Crowding out their purple brethren, blue-greens swarmed the photic zone, the region illuminated and irradiated by the sun and extending to no more than two hundred meters below the ocean surface. In warmer months cyanobacteria grew most vigorously. Sliming along surfaces, trapping and binding sediment, they formed reefs along the coast and damp, textile-like masses inland.

Rust, iron oxide, formed as bacterial waste oxygen reacted with dissolved iron. Since iron is Earth's fifth most abundant element, iron oxides formed in great profusion and, as solids, settled quietly, year after year, to the bottoms of lakes and seas, including basins recently hollowed out by meteorites. Flourishing in the heat and growing more slowly in the cool, cyanobacteria produced more or less oxygen in accordance with seasonal and extended climate changes. Environmental oscillations led to an alternation between oxygen-rich and oxygen-poor varieties of iron ore; magnetite is less oxidized and hematite more so. The seasonal and climatic shifts in population growth, metabolic activity, and community structure of oxygen-producing bacteria, coupled with changing environmental conditions, formed great banded rocks. In North America banded iron formations in ancient, exposed rocks now stretch from eastern Ontario to the western edge of Lake Superior. [PLATE 25] They are the source of the iron used in Detroit-built cars.

Two thousand million years ago not only iron but sulfur, uranium, and manganese oxidized all over the world. Exposed to bacterial waste gas, they became hematite, pyrite, uraninite, and manganese dioxide. As atmospheric oxygen increased, banded iron, which harbors nonoxygenated minerals, gave

25
Alternating bands of hematite and magnetite, banded iron formations (BIF) are a source for much of the world's iron tools and machines. Such banded iron may have been formed by cyanobacteria growing in seasonal cycles of bloom and retrenchment. Ultimately so much oxygen was introduced by bacteria into Earth's atmosphere that iron precipitated not in bands but in rusty heaps called "red beds."

way to "red beds"—rust formations that formed all over the world. The rock record of oxidized minerals in Earth's crust testifies to the addition of oxygen to our planetary atmosphere in a 400-million-year stretch from 2,200 to 1,800 million years ago. Eventually no more minerals were left that had not already reacted with oxygen, so the excess gas with no place left to go began to accumulate in the air.

Quintessential Polluters, Quintessential Recyclers

In what passes for humility and respect for the ways of nature, modern humans worry about our pollution of Earth. Pollution is certainly distressing. But it is hardly unnatural. The pollution crisis effected by all-natural, blue-green bacteria was much worse than any we have seen lately. It destabilized the planetary environment. It made Earth inflammable, and to this day only the ancient surfeit of oxygen permits us to strike a match to make fire.

Human industry has increased the concentration of ozone-unfriendly chlorofluorocarbons in the atmosphere some one hundred times, up to about a billionth of a percent. This degree of change cannot even begin to compare with the effect upon the global environment wrought by the blue-greens. By growing, they increased atmospheric oxygen concentration from less than one part in 100,000,000,000 to one part in five (20%). And Earth's protective, ultraviolet-shielding layer of ozone (O_3, a three-oxygen molecule) was built up largely by "all-natural" pollution in the first place.

But if pollution is natural, so is recycling. Our fresh air is one-fifth oxygen. Today the ozone layer protects animals such as ourselves from ultraviolet skin cancer, cataracts, and compromised immune systems. One of the greatest turnarounds in evolution was the transformation of a once-fatal form of air pollution—oxygen—into a coveted resource.

Far from destroying the planet, oxygen energized it. In far-from-equilibrium systems waste products necessarily accumulate. But what may be garbage to one is dinner or building materials for another.

Bacteria, the greatest metabolic innovators, are not only the greatest polluters but the greatest clean-er-uppers. Our own chemical ability to use oxygen for energy derives from bacteria. Natural pollution recycling by bacteria extends to a host of other substances. Green and purple sulfur bacteria, starting with sulfide, produce sulfur globules and sulfate (both are more oxidized forms of sulfur), which suspend or dissolve in sea water. This sulfur is taken up and recycled by fermenting, sulfate reducing, or even other photosynthetic types of being.

Bacteria, in another one of their global mega-tricks, take nitrogen gas lost to the air and return it to all other living beings, where it is essential for the construction of proteins. Only a few types of bacteria own this miniaturized industry, as only a few are capable of breaking the strong triple bonds of molecular nitrogen and then sequestering the nitrogen atoms into organic molecules without oxygen sneaking in somewhere along the way. Bacteria thus "fix" gaseous nitrogen — by far the most abundant gas in the atmosphere — into organic compounds for all the living beings on Earth. Nitrogen-fixing structures, called heterocysts, were left 2,200 million years ago in the fossil record. These are large cells in chains mainly made of smaller ones. Cyanobacteria with heterocysts can fix N_2 gas and make it available as food. [PLATE 26]

The creative recycling metabolism of bacteria, combined with the imperative of autopoiesis, insures the biospheric flow of nitrogen, sulfur, carbon, and other compounds. Once nitrogen, for example, is fixed into protein and nucleic acid inside bacterial heterocysts, and once these proteins make their way through the food chain (degraded to amino acids and variously rebuilt along the way, with some leakage into the atmosphere as waste),

26
Fischerella, cyanobacterium. Phylum: Cyanobacteria. Kingdom: Monera. An example of bacterial metabolic "superiority," this cyanobacterium fixes atmospheric nitrogen in its heterocysts, making protein. The biochemical and metabolic repertoire of bacteria make them crucial to biological functioning on a global scale.

bacteria are summoned once again to do what only they can do: fix nitrogen back into organic molecules. The organically bound nitrogen in proteins and amino acids takes many routes. Some is degraded to ammonia (NH_3) by a diversity of bacteria. Ammonia is oxidized to nitrite (NO_2) or nitrate (NO_3) by still other bacterial specialists. Nitrite and nitrate, in turn, fertilize the water letting cyanobacteria and others grow. Nitrite and nitrate may be "breathed" by some bacteria which vent nitrous oxide ("laughing gas") and nitrogen (N_2) into the air. Nitrogen gas in the atmosphere must then be fixed again. The complex cycle never ceases. Although no bacteria yet degrade the refractory carbon-hydrogen compounds of most plastics, eventually some will evolve and, not limited by food supply, they will spread like wildfire from landfill to landfill through the biosphere.

Living Carpets and Growing Stones

Like magic carpets in certain remote corners of Earth "microbial mats"—huge numbers of interliving bacteria—have the power to take scientists back in time. [PLATE 27] Slick and slimy, often stinking of sulfur, microbial mats preserve the primeval scene of the early Earth before oxygen surfeited the air. The moist, multicolored mats found just inland from the sea feel cool beneath bare feet. Save for a sheet of algae, sand fly egg, occasional gull overhead, or paleobiologist footprint, nonbacterial traces are rare in microbial mats.

Modern microbial mats and scums are found everywhere in the world, but only in a few locations are they not obscured by larger forms of life. At Laguna Figueroa and Guerrero Negro in Mexico's Baja California, fronting the coastal town of Beaufort in the state of North Carolina, at Plum Island and Sippewissett in Massachusetts, and along the edge of the great Salt Lake in Utah and the sprawling Ebro river in Spain, mats are conspicuous. Visible in places too hot, cold, windy, or salty to

support larger life, microbial mats represent what life on a bacterial planet might have looked like several thousand million years ago.

Merging metabolic talents, a variety of bacterial forms organized into layers thrive in the mats that they themselves produce. The sun-loving cyanobacteria dwell in the upper layers, subtly and continuously transforming carbon, nitrogen, sulfur, and phosphorus—supplying these to their dependents below. Most are multicellular gliding filaments. Some are single-celled. Together, shaped as threads, spheres, or branches, the cyano colonies may form green jelly balls or scums in the water. Among the photosynthesizers, the most light-tolerant, desiccation-resistant forms inhabit the upper sheets of microbial mats; the sulfide-using and dim light-seeking members cohabit the lower layers. The purple sulfur bacteria navigate the middle realm, balancing a need for sulfide from the nether regions with a requirement for sunlight from the higher ones.

Lung-like, not only do the gases in the bacterial community move up and down daily, but so do the constituent guilds. Yehuda Cohen and his colleagues at the marine station at the Gulf of Eilat in Israel have determined that when the sun sets the purple layer of sulfide bacteria moves up a fraction of a centimeter. Deprived of sunlight, the blue-greens above no longer photosynthesize. When again at daylight the blue-greens begin bathing the bacteria below with their oxygen waste, that purple layer retreats.

Like a troop of sea turtles on the march—their shells jutting out of the shallow ocean in Shark's Bay, Australia—are the domed, layered rocks called stromatolites. [PLATE 28] The American geologist Charles Walcott called the fossil remains of very similar but very ancient rocks in Albany, New York, "cryptozoans" (from Greek for "hidden animals"). The locals in Saratoga Springs and around Skidmore College today call these formations "cauliflower

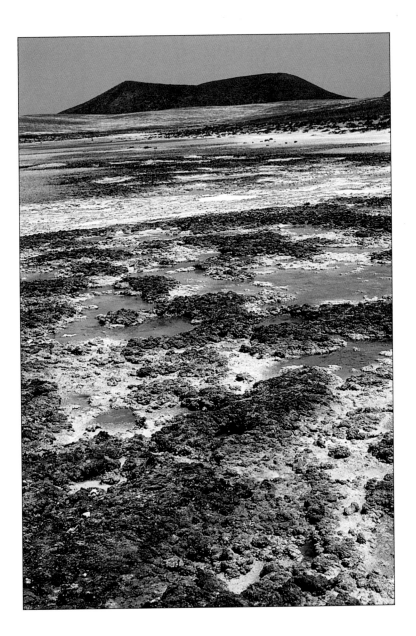

27
What appears as sludge in this beach-side lagoon is actually tightly packed bacterial communities. The upper layers grow photo-synthetically, providing excess which feeds other bacteria in metabolically sophisticated ecosystems, once dominant on Earth, but now confined to areas too harsh, hot and salty for larger life forms. The microbial mat in this picture comes from Laguna Figueroa in Baja California Norte, Mexico.

28
Living bacterial stones called stromatolites from Shark Bay, Australia. Over time some types of microbial mats are thought to give rise to these strange domes. The photograph displays live stromatolites, replete with growing bacteria. Such structures are known from sea-side locations around the world. They occur in both fossil and live form. Bacterial land scapes were common before the evolution of animals, fungi or plants.

29

A petrified fossil stromatolite from Warrawoona,
South Africa (top) is compared with a cut section of
a living microbial mat,
Matanzas, Cuba (bottom).

limestone." Although Walcott had an inkling that the round limestone "heads" were produced by life, only recently has it become clear that "cryptozoans" are stromatolites made by vast hordes of bacteria. They are, in essence, fossilized microbial mats that took the form of domes rather than the columns, reefs, and pancakes typical of the more spectacular ancient stromatolites. [PLATE 29]

The microbial communities, led by blue-greens, trapped, precipitated, and bound up calcium carbonate and grains of volcanic glass before they died. In Australia, where stromatolites are forming today by the work of live bacterial communities, the trapping and binding can be studied directly. Stromatolites (which sometimes make use of environmental silica or even iron rather than only carbonate for building material) grow a layer at a time, as photosynthesizing bacteria glide past one another, slipping out of their polysaccharide sheaths, which are carbohydrate encasements similar in chemical composition to mucus. The sheaths are sticky and they bind sand. The live cyanobacteria, gliding toward the sun, leave behind their sheaths to be colonized by other shelter-seeking microbes. Trapping sediment and precipitating carbonate out of water, some of these complex mat communities solidify to make living fortresses against spume and wave. These fortresses thrive, as many kinds of photosynthetic bacteria support a plethora of camp followers. Spirilla, spirochetes, coccoids, and spore-formers in lively communities jostle for space, food, and position.

Some fossil stromatolites (such as those in the rocks of the Pongola Group in Africa, the Warrawoona Group in the Pilbara region of Western Australia, and the Swaziland in South Africa) harbor microscopic bacterial imprints. These silica stromatolites, made of black chert, are noteworthy for containing the microfossils that provide the best evidence for life in the Archean eon.

Thus, among their other accomplishments, bacteria created hard structures two thousand million years before the first animal evolved. Stromatolite hillocks would have been a common scene in the late Archean eon. Like miniature cathedrals, they were an early manifestation of life's ability to manage its excess. A landscape similar to that seen at Shark Bay has been in continuous existence somewhere on Earth since life began.

On a global scale the living tissues made by microbial mats—whether as living carpet or growing stone—may be as important to biospheric functioning as lung and liver are to us. Bacteria took over the world and still run it, using their decentralized planetary metabolism and capacity for worldwide intraspecies gene transfer.

So, what is life?

Life is bacterial and those organisms that are not bacteria have evolved from organisms that are. By the end of the Archean eon every desert was encrusted with microbial mats and temporary scums; every hot pool, sulfurous or ammoniacal, boasted hoards of colonists and pushy immigrants. Over salt grains and in rusty pools bacteria fabricated glues and precipitated magnetite. Clinging to the cold, barren rocks near the poles and sliming over the volcanic rubble in the tropical shallow seas, greening the Earth, photosynthesizers exuded their wares to hungry opportunists. The waste of a fermenter became the food of the acid-loving swimmer, while the fetid breath of a sulfate reducer provided a precious raw material to green chlorobia or red chromatia. Every available piece of real estate on this planet was occupied by enlightened producer, busy transformer, or arctic explorer. Naturally selected offspring survived, but only if lent a plasmid-borne gene from a community member. Gene exchanges were indispensable to those which would rid themselves of environmental toxins: a protein to be degraded, a poisonous manganese scum, or a threatening copper sheen to be oxidized or reduced. Replicating gene-carrying plasmids owned by the biosphere at large, when borrowed and returned by bacterial metabolic geniuses, alleviated most local environmental dangers, provided said plasmids could temporarily be incorporated into the cells of the threatened bacteria. The tiny bodies of the planetary patina spread to every reach, all microbes reproducing too rapidly for all offspring to survive in any finite universe. Undercover and unwitnessed, life back then was the prodigious progeny of bacteria. It still is. ■

5
Permanent Mergers

I have also seen a sort of animalcule that had the figure of the river eels: These were in very great plenty, and so small withal that I deemed 500 or 600 of them laid out end to end would not reach to the length of the full grown eel such as there are in vinegar. These had a very nimble motion, and bent their bodies serpentwise, and shot through the stuff as quick as a pike does through water.

—Anton van Leeuwenhoek, 1681

We cannot fathom the marvelous complexity of an organic being; but on the hypothesis here advanced this complexity is much increased. Each living creature must be looked at as a microcosm—a little universe, formed of a host of self-propagating organisms, inconceivably minute and as numerous as the stars in heaven.

—Charles Darwin, 1868

The greatest division is not even between plants and animals, but *within* the once-ignored microorganisms—the prokaryotic Monera and the eukaryotic Protoctista.

—Stephen Jay Gould, 1982

The appearance of these [protoctist] cells a billion-odd years ago was the second major event in planetary evolution and led directly, lineage by lineage, to our own complex selves, brain and all.

—Lewis Thomas, 1990

The Great Cell Divide

Some 2,000 million years ago, probably at many different sites on Earth, a new kind of cell evolved from bacterial interactions. The evolution of these complex new cells from integration of bacterial symbionts prepared the way for life in the new, Proterozoic eon. These new cells were ultimately the result of hunger, crowding, and thirst among teeming bacteria. These new cells were the first protoctists, and their coming brought the kinds of individuality and cell organization, the kind of sex, and even the kind of mortality (programmed death of the individual) familiar to us as animals. Bacteria merged. Curbing their viciousness and surrendering independence, they explored new ways to persist and reproduce.

Our kind of life, that of the nucleated cell, began long before animals. Amid cell gorgings and aborted invasions, merged beings that infected one another were reinvigorated by the incorporation of their permanent "disease." The first new kind of cell—the nucleated cell—evolved by acquisition, not of inherited characteristics but of inherited bacterial symbionts. These new kinds of cells making up the bodies of unicellular protists and multicellular protoctists would eventually lead to the final three kingdoms of life yet to evolve on Earth: animals, fungi, and plants. Our protoctistan ancestors were beings so exceedingly weird that, if informed in detail of their existence, even the credulous author of a medieval bestiary might pooh-pooh the tale as the impossible product of a febrile imagination.

Each and every organic being on Earth is made of one of only two kinds of cells. Our kind—and that of other animals, fungi, plants, and protists—possesses nuclei. The other kind, the bacterial cell, has no nucleus. In 1937 Edouard Chatton, a French marine biologist, named the latter cell type "procariotique"; the organisms that possess this cell type are prokaryotes (pronounced, "pro-CARRY-oats"). All the rest of us are eukaryotes ("you-CARRY-oats"), made of nucleated cells. So the presence of a membrane-bounded nucleus defines a cell as "eukaryotic." All eukaryotes come from protoctists; bacteria don't. The long DNA molecules, the genes of eukaryotes, are organized within the nucleus into at least two and as many as several thousand chromosomes (humans have forty-six). As we shall see, this sequestering of precious genetic material inside a special membrane and the firm binding of DNA into a particular sequence in a particular chromosome limited the genetic promiscuity that was and still is accepted practice in the bacterial realm. [PLATE 30, *following page*]

A giraffe is a eukaryotic organism, made of eukaryotic cells. So is a daisy. And an ameba. The differences in behavior, genetics, organization, metabolism, and especially structure between prokaryotes and eukaryotes are far more dramatic than any between plants and animals. Those differences mark the great cell divide. Prokaryotes and eukaryotes thus form the two "supergroups" of life on Earth.

All of one supergroup and a good portion of the other inhabit the microbial realm. Bacteria, the smaller protoctists, and yeasts and other small fungi are microbes. The eukaryotic cells of protoctists and fungi are bigger than the prokaryotic cells of bacteria; but, like any cell, they must be viewed with a microscope. The path of transition between the two supergroups is obscure. The evolution from prokaryotes to eukaryotes, from bacteria to protoctists, was a "symmetry break" that catapulted life to a greater level of complexity and gave it different potentials and risks. Not just by gradual mutation but suddenly through symbiotic alliance did the first eukaryotes form.

Five Kinds of Beings

The earliest eukaryotic cells, living on their own, were protoctists that evolved by permanent bacterial merging. Floating or free-swimming, some went on to become animals, fungi, and plants.

The protoctists are a wide-ranging group of obscure beings. Today an estimated 250,000 species include tiny amebas and diatoms and giant kelps and red seaweeds. Ultimately, this group gave rise to familiar plants and animals such as palm trees and clams. But even as recently as a thousand million years ago not a single animal, plant, or even fungus dwelled on Earth. Biospheric functions were handled entirely by bacteria and protoctists.

The ungainly name "protoctist" was introduced by an English naturalist with the equally unenviable name: Hogg. John Hogg (1800-1861) set forth his views in an article published in 1861, just before he died: "On the Distinctions of a Plant and an Animal, and on a Fourth Kingdom of Nature."[1] (His third was the "mineral kingdom.") Neither Hogg nor anyone else at that time was aware of prokaryotic and eukaryotic cells. But Hogg saw that many organisms were neither plant nor animal.

Unlike the term protozoa ("first animals"), with its unfortunate connotation that organisms ranging from foraminifera to slime nets were somehow animals, protoctist simply means "first beings." Protoctists are neither animals nor necessarily single-celled. But when they are single-celled—or otherwise tiny—they are called protists. Because all animals grow from multicelled embryos, there are, by definition, no single-celled animals. So-called single-celled animals are really the protists, the smaller protoctists. Hogg suggested "Regnum Primogenium" as the name for this primordial kingdom. Its founding members are now known to have originated prior to plants and

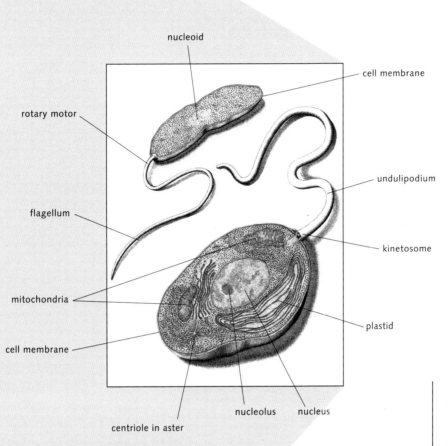

nucleoid

cell membrane

rotary motor

undulipodium

flagellum

kinetosome

mitochondria

plastid

cell membrane

centriole in aster

nucleolus   nucleus

30
Illustrated comparison of a prokaryote—bacterium— at top and a eukaryote— nucleated cell—at bottom. All living cells on Earth are either prokaryotes or eukaryotes. The non-bacterial kingdoms— Protoctista, Fungi, Plantae and Animalia—all consist of organisms whose cells are eukaryotic. Eukaryotes evolved symbiotically from eating, invading, infecting, and cohabitating bacteria.

animals, and yet protoctists continue to thrive on Earth today. [PLATE 31]

In Germany Ernst Haeckel also argued for a new kingdom. "These interesting and important organic beings are the *primary creatures* or *Protista*."[2] The Monera—bacteria—were part of Haeckel's proposed Protista. Haeckel, recall, was not persuaded by Lazzaro Spallanzani's boiling of mutton broth to kill microbes. It seemed clear to him that primordial beings simpler than anything yet discovered must exist. An ardent believer both in evolution and in the spontaneous generation of matter, Haeckel sought "an entirely homogeneous and structureless substance, a living particle of albumin, capable of nourishment and reproduction."[3]

English biologist Thomas Henry Huxley (1825-1895) was taken with Haeckel's notion of a primordial protein globule. Examining ten-year-old mud samples dredged up from the seabed off the northwest coast of Ireland, Huxley discovered a mysterious white ooze. Were these Haeckel's postulated earliest Protista? Upon examination, the granular ooze was seen to consist of tiny calcareous plates. Excited, Huxley wrote to Haeckel that he had encountered the ancestral life form. Indeed in the flush of his discovery, Huxley honored his colleague by naming the "organisms" after Haeckel. Both men delivered the exciting news that *Bathybius haeckelii*, the great Urschleim (primordial goop), had finally been found.

Only later was it realized that *Bathybius haeckelii* was just marine sediment. The white slime that appeared whenever Huxley doused the ooze to preserve it was an alcohol precipitate of organic debris that included jellyfish stingers. Far from being our primordial parent, the Urschleim was not even alive. Nevertheless, Haeckel's concept focused scientific attention on beings that escaped the dichotomous plant/animal classification scheme.

Today the tendency to divide life into animal versus plant remains. Fungi, if they exist at all in the

31
Mesodinium rubrum,
a protist. Phylum:
Ciliophora. Kingdom:
Protoctista. This fast-swimming but photosynthetic microscopic being dwells in brackish water near the Baltic Sea. It exemplifies organisms confounding older, three-kingdom classification systems since it is really neither plant nor animal. The reddish hue of the interior derives from degenerate symbiotic algae.

popular imagination, are a kind of gray plant. Smaller protists and bacteria—not quite life in the popular mind—are ignored or lumped together as "germs." Academia still departmentalizes life into botany, the study of plants, and zoology, the study of animals. Fungi, bacteria, and certain protoctists are often forced in this scheme to be plants under the jurisdiction of botanists. This quaint plant-animal split does not reflect evolution. The ancestors to plants and animals were neither; rather they were communities—bacteria that merged to form a new kind of cell.

The first essentially modern classification was invented by Herbert F. Copeland (1902-1968), a biology teacher at Sacramento City College in California. Copeland argued for four kingdoms: Monera (bacteria), plants, animals, and protoctists. He placed all fungi (molds, mushrooms, puffballs, etc., which he called "Inophyta") into a subdivision of Hogg's Protoctista. His book, *The Classification of the Lower Organisms*, published at Copeland's own expense by a so-called vanity press, was read by almost no one except Cornell University ecologist Robert H. Whittaker (1924-1980). Whittaker devised the most useful groupings of all when he removed fungi from his Protista and recognized them as a distinct "fifth kingdom."

From today's vantage, Whittaker's five-kingdom classification scheme best reflects evolutionary relationships. One of us (Lynn Margulis) has collaborated with zoologist Karlene Schwartz of the University of Massachusetts at Boston to sharpen the blurred boundaries of Whittaker's protists. The Kingdom Protoctista, which Whittaker limited to unicellular and the smallest multicellular beings, now includes larger organisms that are not plant, animal, fungi, or bacteria, such as seaweeds.

Twists in the Tree of Life

The story of how a human—a being made of nucleated cells—evolves from an ameboid being—a nucleated cell—is bizarre. But even this story has a preamble: the evolution of a cell with a nucleus. How did such a cell evolve?

The quick answer is by the merging of different kinds of bacteria. Protoctists evolved through symbiosis; twigs and limbs on the tree of life not only branched out but grew together and fused. Symbiosis refers to an ecological and physical relationship between two kinds of organisms that is far more intimate than most organism associations. In Africa, for example, plovers pluck and eat leeches from the open mouths of crocodiles without fear. Bird and beast in this instance are behavioral symbionts; crocodiles enjoy clean teeth in the company of well-fed plovers. Bacteria live in the spaces between our teeth and in our intestines, mites inhabit our eyelashes; all these tiny beings draw nutriment from our cells or our uneaten food, as cells are shed or as they excrete organic excess. Symbiosis, like marriage, means living together for better or worse; but whereas marriage is between two different people, symbiosis is between two or more different types of live beings.

Organisms form many kinds of symbioses, but the most awe-inspiring is the exceedingly close association known as endosymbiosis. This is a relationship in which one being—microbe or larger—lives not just near (nor even permanently on) another, but inside it. In endosymbiosis, organic beings merge. Endosymbiosis is like a long-lasting sexual encounter except that the participants are members of different species. Indeed, some endosymbiotic linkages have become permanent.

Bacteria, masters of symbiosis in general, are also the best endosymbionts for at least four reasons. First, because they have been entering into stable relationships with one another for several thousands of millions of years, they are good at forming permanent relationships. Second, their tiny bodies fluidly lose and acquire genes, making them amenable to rapid genetic change. Third, bacteria have only a limited expression of individuality; no circulating antibodies guard them — an "infection," far from being rejected as it might be in an animal with immune system, can thus become the basis for life-long association, a mutual evolution. Fourth, bacteria's vast chemical repertoire leads to a tendency for metabolic complementarity less often seen in associations between already highly individualized members of plant and animals species. Of course, given time, some plants and animals may come together as closely as some bacteria have.

Symbiosis produces new individuals. "We" could not synthesize B or K vitamins without bacteria in our gut. Cows and termites are not themselves without the swimming fermenters in their digestive systems — ciliates and bacteria that break down grass and wood. Some algae living inside translucent flatworms are such good providers that the worms have atrophied mouths; the close-mouthed green worms "sunbathe" rather than seek food, and the endosymbiotic algae even recycle the worm's uric acid waste into food.

Thousands of other strange partnerships exist. All of the estimated 20,000 lichens, for example, began as symbiotic associations of algae with fungi or of cyanobacteria with fungi. But the most important symbioses were those that led to the eukaryotic cell.

Today most protoctist cells and all plant, animal, and fungal cells contain mitochondria. The oxygen-respiration that keeps members of the youngest four kingdoms of life alive takes place inside these particular organelles. (Like organs within bodies, organelles are functioning structures within eukaryotic cells.) Mitochondrial organelles look like bacteria. They even grow and divide in two at their own pace within the larger cell. They are thought to come from bacteria — but after more than a thousand million years of association they cannot survive outside the confines of the cell.

The cells of plants and some protoctists such as algae also possess colorful bodies called plastids. All the photosynthesis undergone by algae and plants happens inside the DNA-containing plastid organelles. Plastids contain the same pigments and other biochemicals found in the spherical, oxygen-producing blue-green bacteria that thrive in the ocean. Coincidence? We don't think so. Indeed, DNA in the plastids of the cells of the red seaweed *Porphyridium* is closer in its nucleotide sequence to that of certain cyanobacteria than it is to DNA in the nucleus of the red seaweed itself.

Such genetic evidence links the cell organelles to their origin from free-living bacteria in a definitive (and now virtually undisputed) way. Genetic similarities that cross kingdoms is the biological equivalent of ancient "fingerprints," proving that photosynthetic organelles did not evolve gradually by a buildup of mutations in the DNA of plant and algal progenitors, but suddenly when symbiotic bacteria took up residence in larger cells. In a moment we will return to the question of how the bacteria that became mitochondria and plastids found their way to their current, cozy location inside the cell. But, to be chronologically correct, we must first explore what may be a still older, and deeper, symbiosis.

Squirmers

Nearly all biologists now accept that particular bacteria, after a period of chemical negotiation and gene transfer, began as symbionts and became the mitochondria and plastids of larger cells. Most biologists, however, reject or are ignorant of another idea. Nonetheless, circumstantial evidence suggests that a still older bacterial symbiosis preceded the acquisition of these organelles.

Before oxygen-users infected anaerobic, swimming protists to form alliances, and before blue-green bacteria were engulfed by these alliances, faster bacteria seem to have conjoined.

32

Trichonympha, a chimerical protist. Phylum: Zoomastigina. Kingdom: Protoctista. This being, as peculiar structurally as any to be found in a medieval bestiary, is composed of a large protoctist host and a swarm of both undulipodia (its organelles in the front) and symbiotically attached spirochete bacteria in the rear. It is itself symbiotic in the termites' hindgut, a microscopic zoo containing many different sorts of protists and bacteria that together aid in the digestion of wood.

Transforming from free-living bacteria to become parts of cells, wriggling spirochetes may have conferred their considerable powers of movement upon the outside, and then the inside, of victims that became ancestral cells. Today's spirochetes are proton-powered bacteria that ferment carbohydrates and whip about like possessed corkscrews. The most rapid swimmers of the entire moneran kingdom, they literally screw their way through mud, tissue, and slime.

Thriving in saliva, the crystalline-styles—digestive tissues—of oysters, the hindguts of termites, and a thousand other equally ingenious niches, invasive spirochetes are one of the most successful life forms on Earth. And they form alliances—often reversibly attaching to larger organisms and propelling them along. Some protist cells—such as *Mixotricha paradoxa* and *Trichonympha*—have gone so far as to evolve holdfast structures where free-living spirochetes are encouraged to reversibly "dock"—with their engines running. [PLATE 32] The spirochetes actively feed on the metabolic leftovers of the cells to which they attach. The symbiotic advantage is obvious: the squirming spirochetes move the cells that feed them. A *Mixotricha* or *Trichonympha* cell without spirochetes is like a boat lacking a motor or a teenager without a car. Consortia able to swim quickly have more opportunities than their sluggish predecessors to find food, escape predators, and meet mates. But external spirochetal attachments are not the whole story. [PLATE 33]

33

In this drawing artist Christie Lyons portrays spirochetes attaching and becoming the undulipodia of larger, new eukaryotic cells.

bacteria

protist

free-swimming spirochete

protist

spirochetes becoming undulipodia

undulipodium

bacterium

34

Different stages of mitosis, the usual method of chromosomal separation prior to cell division in reproducing eukaryotic cells. The greater internal movement of cells with nuclei as compared to bacteria may result from the 2,000 million-year-old remnants of rapidly squirming spirochetes.

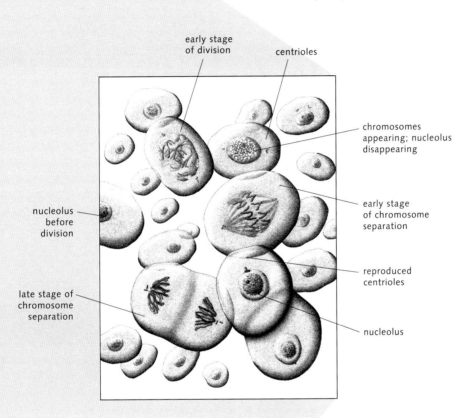

early stage of division

centrioles

chromosomes appearing; nucleolus disappearing

early stage of chromosome separation

nucleolus before division

reproduced centrioles

late stage of chromosome separation

nucleolus

Protoctist cells, huge compared to bacteria, display incessant internal movement. Bacterial cells, lacking any internal movement and real chromosomes, do not divide mitotically; they do not perform "the dance of the chromosomes." Mitosis, the chromosome style of cell reproduction, is widespread among protoctists, and universal in their animal, plant, and fungal descendants. Matching chromosomes line up and move to opposite poles in a kind of micro-ballet. At the mitotic poles in animal and many protist cells are centrioles, structures resembling rotary telephone dials that may be the remnants of spirochetes that long ago entered larger cells to feed. [PLATE 34]

The mitosis by which most eukaryotic cells divide ensures that chromosomes doubled in the parent are

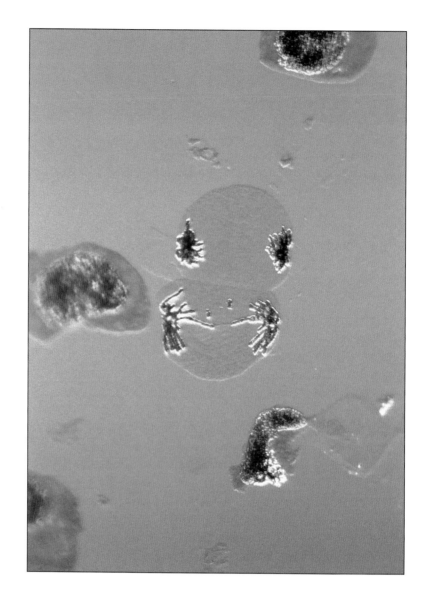

35
Haemanthus sp., African
blood lily. Phylum:
Angiospermophyta.
Kingdom: Plantae.
Telophase in the mitotic
cells deep inside the
flower. Mitosis is the
kind of reproduction of
cells by division that typi-
fies the nucleated cells of
all animals, plants, and
fungi and most protoc-
tists. Telophase is a late
stage of mitosis. In mitosis the chromosomes dou-
ble first and then divide
and separate into these
two masses that become
the nuclei of the resulting
offspring cells. In the
mitotic cell division
shown here the flower's
chromosomes are stained
red.

partitioned evenly into two offspring cells. Mitosis seems indispensable as a genetic filing and distribution system for the huge quantities of DNA that most eukaryotic cells contain. In each episode of mitosis a series of tiny protein tubes, microtubules (collectively called the mitotic spindle), appears. At the end of the process, when one cell has become two, this mitotic spindle disappears. The chromosomes attached to the tubules of the spindle line up along the plane of the cell's equator. These chromosomes, which doubled earlier, now separate as each half moves along the spindle to the opposite side of the cell. The chromosomes now at the poles uncoil as the cell proceeds to divide into two. The enigmatic mitotic spindle then fades back into the invisibility from which it emerged. [PLATE 35]

In many animals the centrioles (the telephone dial-like structures) move to the edges of the cell, where they become kinetosomes by growing shafts. In cross section, the shafts show a distinctive "9(2)+2" pattern: nine sets of two tubules arranged near the perimeter of the circular axis, with one set of two tubules at the center. The kinetosome and centriole have different names only to distinguish the shafted from the shaft-less phase of the same organelle. Multiple names for the same organelle are more accidents of history than nomenclatural necessities. Two names were bestowed because the distinct phases were noticed and named long before any relationship between them was recognized.

The universality of the kinetosome shafts—found in plants, animals, fungi, and protists—is strong evidence for an ancient origin. The 9(2)+2 symmetry is found, for example, in the cell extensions of the balance organ of our inner ears and in tails propelling the swimming protist *Euglena*. The 9(2)+2 arrangement can be seen in cross section in sperm cells of men. Because of their similarity, all 9(2)+2 shafts that grow from kinetosomes are best referred to by a common name. We call them undulipodia ("waving feet").

The cell reproduction of mitosis—remarkably similar in some protoctist and all plant, animal, and fungal cells—must have evolved in the oldest of these four kingdoms. The protoctists, from which emerged plants, animals and fungi, were the first beings with this motility apparatus necessary for the reproduction of the new cells with nuclei. But it seems doubtful that the smaller protoctists invented the undulipodium and internal cell movement. Motility, rather, could have been the gift of the oldest and original kingdom of life.

Tantalizing evidence implicates a bacterial origin of the centriole-kinetosome organelle. Both DNA and RNA have been reported in these intracellular structures. David Luck and John Hall at Rockefeller University in New York City photographed a peculiar, bacteria-like DNA in the two centriole-kinetosomes of the green alga *Chlamydomonas*. Joel Rosenbaum and his colleagues at Yale University, along with several other scientists working independently, have not, however, been able to confirm any centriole-kinetosomal DNA in this green alga.

Living cells are decorated by undulipodia that go by many different names. Undulipodia include all cilia and the "tails" of most sperm. The single-tailed swimmer in bull semen and the hundred-tailed sperm released by male fern plants are both examples of 9(2)+2 undulipodia. The immotile cilia remnants in the rod and cone cells of our retinas, the motile ones of the fallopian tube cells that push a woman's egg toward the womb, and those rejecting debris in our windpipes are further examples of undulipodia.

It may be that the spirochetes that symbiotically became undulipodia (involved both as cell tails and chromosome movers) have become so integrated with their partners that they have dissolved away to mere traces and genetic shadows of their former selves. Like an artist who seems effortless in the performance of a difficult routine, the former spirochete genes may be so deeply implicated in cell function that today they all but defy detection. Oxford University biologist David C. Smith likens such symbiotic remains to the smile of the Cheshire Cat, the fictional feline of Lewis Carrol in *Alice in Wonderland* who slowly fades away to become nothing but an enigmatic grin, floating in midair: "the organism progressively loses pieces of itself, slowly blending into the general background, its former existence betrayed by some relic."[4]

The remaining traces of the bequeathers of motility are fewer and hazier than those left by the cells which gave the green gift of photosynthesis and oxygen-bubbling respiration. Motility, in our view, was the first endosymbiotic acquisition of the nascent eukaryote; losing parts of themselves as they evolved, squirming spirochetes invaded and animated what were to become nucleated cells. Today, because of time, the evidence is thin. The photographs have all but faded and the pages have crumbled. Cell history must be reconstructed from the faintest clues.

A reason to think that spirochete symbiosis preceded the others is the recent discovery of many protists that have undulipodia but lack mitochondria. These air-shunning beings are poisoned by oxygen—suggesting they date from a time before ancestral protists had become symbiotic with the oxygen-using bacteria that evolved into mitochondria. Mitotic cell division in which chromosomes line up on the spindle is universal in animal, plant, and fungal cells. Only a few dark-dwelling, oxygen-shunning, swimming "amitochondriate" protists and their obscure relatives show important variation on the mitosis theme.

The absence of intermediates between bacteria and such seemingly aberrant protists tells us that evolution from bacteria to nucleated but still anaerobic swimmer probably did not occur by random mutation alone. The sudden evolution of cells with nuclei and 9(2)+2 swimming organelles is best explained by ancient motility symbiosis. When the close connection of undulipodia and mitotic apparatus is observed in live anaerobic cells, symbiosis becomes the most parsimonious of all scientific explanations. Indeed, by comparison, mutation explanations for the origin of undulipodia seem far-fetched.

Consider a very ancient ancestor of one of today's bacterial denizens of hot springs, *Thermoplasma* [see PLATE 35]. Imagine that ancestor under attack by spirochetes. Holding firm, its protective membrane resists penetration. The spirochetes attach on the outside, establishing association, as they feed on *Thermoplasma*'s waste. Eventually some gain entry and merge with the debilitated *Thermoplasma* to become its living oars.

Once inside, the spirochete symbionts extend their motility skills to the internal operations of their would-be victim. A sort of biochemical truce prevails, as both sorts of reproducing partner manage to coexist. The nucleus, acting today as a sort of central genetic government, might have evolved as membrane proliferated to keep the attacking spirochetes from eating out *Thermoplasma*'s DNA. The captive spirochetes, still moving, ultimately became movers of chromosomes. Mitosis evolved. Spirochete attachments became centriole-kinetosomes. Perhaps some of these structures—those that have retained the power to reproduce—still contain DNA.

Whatever the precise scenario for the acquisition of motility and the sometimes respiratory and photosynthetic talents of eukaryotes, symbiosis most assuredly belongs in the narrative. Intimate symbioses were essential to the evolution of cells.

Strange New Fruit

The hypothesis that former spirochetes and thermoplasmas merged to form swimming protists is under investigation now. These merged beings may have been the original members of confederacies of bacteria from which all larger life evolved. But what about other symbiotic bacteria? How did they become involved?

Think back to those blue-green, photosynthetic bacteria that polluted Earth with oxygen gas. After reacting to make new minerals such as sulfate (SO_4), magnetite (Fe_2O_3), and hematite (Fe_3O_4) all over the planet's surface, oxygen waste began to accumulate in the atmosphere. Newly appeared gaseous oxygen killed off untold hordes of organic beings. Even today certain kinds of cyanobacteria are sickened by their own oxygen; *Phormidium*, for example, lives only in muds near other organisms that can quickly use up the oxygen it produces in would-be fatal concentrations.

Early on, cells evolved a tolerance for oxygen at low concentrations. Many modern prokaryotes still function best at oxygen levels of about ten percent— half of the atmosphere's typical concentration today. Oxygen-tolerant bacteria produce enzymes such as catalases, peroxidases, and superoxide dismutases that react with the dangerous gas to produce innocuous organic compounds and water. Without such chemical buffers the carbon of organic tissue is scorched, torched, and laid to waste by oxygen.

Nonetheless, the mitochondria of our cells come from bacteria that neither shunned nor merely tolerated oxygen. The bacteria that evolved into the matrilineally transmitted mitochondria—only the ovum bequeaths them to the human embryo— exploited oxygen's great reactivity. Like nuclear physicists devising a way to power spacecraft by using environmentally hazardous plutonium, the mitochondrial ancestors turned an intense danger into a radical opportunity.

In perhaps the greatest example ever of recycling, bacteria employed reactive oxygen to improve cell processes of energy transformation. Oxidizing the material they produced by trapping the energy of light, purple photosynthetic bacteria increased their ability to metabolize ATP, the energy storage compound—the biochemical "coin" used by every cell of every living being. Breaking down organic molecules and producing carbon dioxide and water, bacteria diverted the natural combustion of oxygen to their own purposes. Whereas on average two molecules of ATP are produced by fermentation of a sugar molecule, with the evolution of respiration the same sugar molecule was made to yield as many as thirty-six ATP molecules. The new bacteria— including the ancestors of our mitochondria— recouped energy from sugar molecules with over fifteen times the efficiency of their oxygen-poisoned predecessors.

That our mitochondria's ancestors were oxygen-respiring purple bacteria has been shown beyond a doubt by DNA sequencing. Like a village ransacked by barbarians that ultimately became civilized, fermenting organisms were attacked by oxygen-using predators that became mitochondrial laborers. We suspect the earliest hosts were *Thermoplasma*-like archaebacteria (already squirming with spirochete symbionts), able to withstand heat and acid but not free oxygen. These consortia evolved into the first protists, their spirochetes becoming undulipodia. The *Thermoplasma* lineage is implicated in this major evolutionary event because modern represen-tatives resemble the nucleocytoplasm portions of eukaryotic cells. *Thermoplasma acidophilum*, for example, possesses histone-like proteins nearly universal in larger life forms—animals, plants, fungi, and some protoctists—but lacking in other prokaryotes. The presence of histone proteins in human chromosomes may be a direct inheritance from the protists that were invaded by proto-mitochondria.

The invaders were probably of the "purple bacterial lineage," as classified by Carl Woese. These proto-mitochondria may have been similar to modern oxygen-respiring, rod-shaped bacteria such as *Paracoccus denitrificans*. This bacterium contains more than forty enzymes in common with human

Bdellovibrio, a predatory bacterium. Phylum: Pseudomonads. Kingdom: Monera. *Bdellovibrio* is caught in the act of invading a larger bacterium. It will multiply inside its prey, devouring away until it breaks the corpse in a living explosion, releasing a new generation of predators. Such invasions when they became permanent, in the sense that the smaller bacteria survive and multiply inside the larger bacteria, could have been analogous to the evolution of cells with mitochondria.

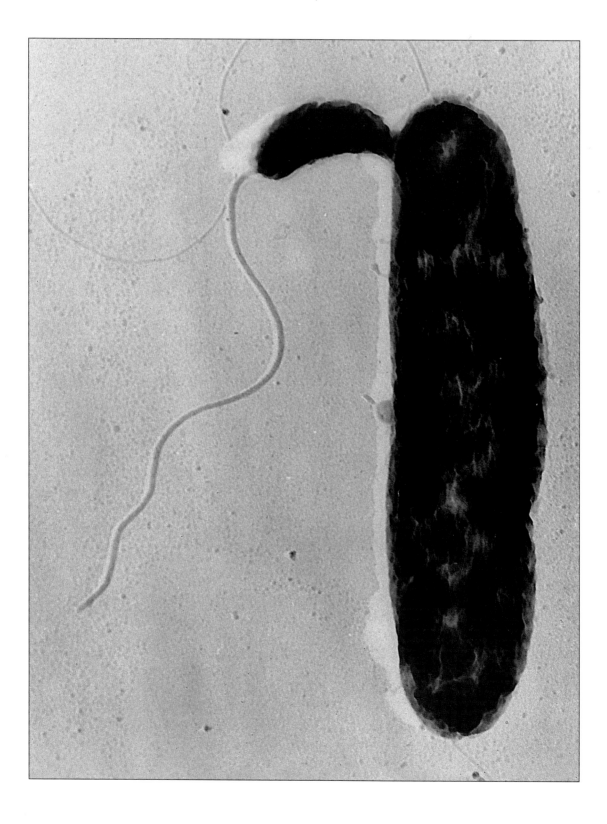

mitochondria. More probably, they were similar to the respiring *Bdellovibrio* or *Daptobacter*—modern predatory prokaryotes in the habit of attacking and multiplying inside larger bacteria. [PLATE 36] Eventually, the victims explode and a battalion of intruders merrily swim out. *Daptobacter*, *Bdellovibrio*, and similar unnamed bacteria are necrobes—beings that live off the death of others. But even if they began as a parasitic infection, the ancestors to the mitochondria did not stay that way. Fed and protected in a living environment, the protomitochondria were better off not destroying their oxygen-intolerant hosts.

Today, although mitochondria still possess their own DNA and still reproduce like bacteria, they cannot live on their own. The parasitism has become permanent: neither partner can escape, neither can survive separation. The first protists were thus odd couples, the results of fusion of two, or (in the case of plants), at least three once-independent beings. But unlike the fire-breathing being of Greek mythology who has the head of a lioness, the midsection of a goat, and a dragon's tail, these chimeras were real.

How do predators become symbionts? How does a deadly infection become a bodily part?

The Korean-American biologist Kwang Jeon at the University of Tennessee has already witnessed such a transformation in the laboratory. The answer is thus less a mystery than before. Jeon's experiments dramatically show how bacteria can change from virulent pathogens to needed organelles.

Like many of science's most amazing discoveries, Jeon's came about accidentally to the prepared mind. To his initial dismay he found one day that his amebas, which he grew in laboratory dishes, were sick and dying. Microscopic investigation revealed that each *Amoeba proteus* was infected with some 150,000 strange bacteria. All but a few amebas died. Curious about the moribund survivors, Jeon injected new, healthy amebas with infectious bacteria

taken from the moribund. Most newly injected amebas died within a few days although, again, some managed to survive. Those that did reproduced more slowly. After some months all the survivors were infected. But these survivors had fewer bacteria inside them than those which had died.

After growing generation upon generation of infected amebas, Jeon extracted the nuclei from several. He transplanted these nuclei into healthy, bacteria-free amebas, whose own nuclei had been microsurgically removed. The amebas with the transplanted nuclei died on the third or fourth day—unless Jeon rescued them with a needleful of bacterial "infection." The disease had thus become the cure. A deadly bacterium had become a vital cell part.

Decades later Jeon's infected amebas are alive and well and living in Knoxville, Tennessee. His experiments have been repeated numerous times, and now he observes that the amebas differ in many features from their never-infected ancestors. Pathogens have become symbionts on at least four occasions. Symbionts have become organelles each time. Invader and invaded merge, evolve into new life forms. Branches on the tree of life do not always diverge but sometimes come together to produce strange new fruit.

Wallin's Symbionts

In 1927 the American biologist Ivan Wallin (1883-1969) wrote, "It is a rather startling proposal that bacteria, the organisms which are popularly associated with disease, may represent the fundamental causative factor in the origin of species."[5] He claimed to have grown mitochondria outside their animal "host cells." Publicly shouted down by colleagues, Wallin gave up defense of the bacterial origin of mitochondria while still in his forties.

Wallin was almost certainly mistaken, as no one has ever been able to grow mitochondria by themselves. Nonetheless, Wallin's theoretical assertions

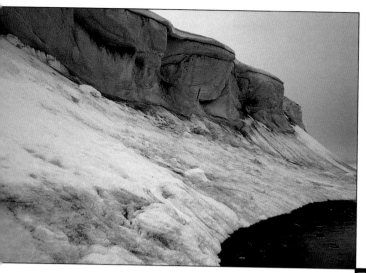

Chlamydomonas nivalis,
an alga. Phylum: Chloro-
phyta (green algae). Kingdom: Protoctista. Red
snow algae in Antarctica.
The green chlorophyll
is masked by the red
pigment. DNA studies of
the green algae and plants
and marine red algae point
to a symbiotic origin of
the colored cell parts from
cyanobacteria. Cyanobac-
teria, teaming up with
larger cells, eventually
evolved into the plastids
of all "higher" photosyn-
thetic beings from
seaweeds to maple trees.

38 (middle)
Medium range photo-
graph of the snow algae
Chloromonas sp. and
filamentous fungi.
The pigments of these
photosynthetic creatures
gives the mountainous
snow its orange tint.

39 (bottom)
Microscopic view of red
snow alga *C. nivalis,* so-
called "watermelon or
candy-colored snow."
The red carotenoid pig-
mentation network serves
as a photoprotectant from
bright sunlight. The cells
are 400 times smaller
than they appear in this
photograph.

were prescient. Plant and animal life, he asserted, had appeared through what he called "symbionticism" or "the formation of microsymbiotic complexes." He meant new species form by the permanent acquisition of symbiotic bacteria.

Today Wallin has been vindicated. His 1927 classic book, *Symbionticism and the Origin of the Species*, was the first systematic description in English of the importance of symbiosis in cell evolution. Although heresy only decades ago, contemporary biologists agree that animals, fungi, and plants evolved from protoctist ancestors themselves originating from symbiotic bacterial associations.

The crucial piece of evidence unavailable to Wallin until just before he died was the discovery that mitochondria and plastids possess their own DNA. Wallin knew though that mitochondria and plastids tend to reproduce at different times than do the cells in which they reside—as if demonstrating a residual impulse of their earlier, wilder days. Bacterial respirers, like those infecting Jeon's amebas, allied with nucleated swimmers to form the ameba-like ancestors of larger life forms: aerobic protists. Combining metabolism and genes, different lineages of aerobic protists went on to evolve into animals and fungi.

Algae and plants are a further chapter in the same story. In subsequent symbiotic events, swimming protists that had already fully integrated with purple bacteria (now mitochondria) came to possess plastids. How? By indigestion. The resistant green bacteria—the food—remained alive inside transparent, vegetarian protists. A continuous supply of photosynthate—food made by the trapped photosynthetic bacteria—rewarded the protist, which quickly developed a penchant for sunlit waters. Like small farmers who harvest their own gardens rather than shop at a grocery, protists which incorporated their captives became increasingly self-sufficient. In return for the favor of food, the engulfed photosynthetic bacteria received a place to live and rapid free transportation into the sunlight.

These swimming protists, which later evolved into algae, were living greenhouses. Would-be food, really endosymbiotic bacteria, photosynthesized inside the luxury prison of live cells. The original undigested food was probably similar to *Prochloron*. This grass-green bacterium grows in the rear chamber—the cloaca—of homely marine creatures known as didemnids or "sea lemons." [see PLATE 9] *Prochloron*-like bacteria are a good scientific choice for the plastids of algal and plant cells. Spherical prochlorons and a rod-shaped (but similarly grass-green) bacterium called *Prochlorothrix* make precisely the same pigments—chlorophylls *a* and *b*—made by green algae and plants.

Multi-tentacled hydras, relatives of jellyfish and coral, are white but tint green when they possess symbiotic green photosynthetic microbes. The snail *Plachobranchus* has gardenlike rows of green plastids under its parapodial folds, part of the digestive tract. The giant clam, *Tridacna*, hosts green dinomastigote algae. Many organisms have allied with photosynthetic bacteria or algae. History repeats itself.

Grass-green and blue-green bacteria are independent versions of the plastids of algal and plant cells. Algal plastids need not be green. Plastids of the alga responsible for the red tint of alpine snow patches in the late spring and summer ("watermelon snow") are red. [PLATES 37, 38, and 39] And in Tanzania's Lake Natrum swoop great flocks of pink flamingos. Red photosynthetic bacteria and algae with red plastids, pigmented with the same carotenoids that color carrots, grow in the lake. Flamingos look pink because the pigments at the microbial base of the food chain wind up coloring the bodies of these intriguing birds.

Genetic evidence, DNA, RNA, and protein sequence information links red algal plastids to certain cyanobacteria with the same forensic accuracy admissible in court to convict a rapist whose DNA matches that of a sperm sample. The multicolored bacteria of the Archean eon have not gone away. They have joined with other cells to become the sea-green chloroplasts of garden cucumbers. Others have become the brown phaeoplasts of kelp in coastal waters. Still others lurk today as the red rhodoplasts of dulse, a form of sea lettuce. If food crops are grown in orbit, on Mars, or on other planets greened with life, it will be a transhuman phenomenon, part of the same bacterial expansion that began more than 3,000 million years ago on the Archean shores.

Multicellularity and Programmed Death

Plants and animals are so complex that it is easy to forget their original status as colonies of hybrids. Occasionally, however, we are reminded of our multicellularity. "HeLa" cells—from the cervix of Henrietta Lane, a woman who lived in Washington, D.C.—continue to be grown in laboratories around the world, despite Lane's death from cancer of that same cervix in the 1950s. This morbid medical fact demonstrates our colonial nature as huge collections of nucleated cells organized into tissues.

By symbiosis different varieties of bacteria came together and made cells with nuclei. These cells with nuclei often cloned themselves into multiple copies that stayed in physical contact after reproduction. A *Paramecium* or *Euglena* is an "individual" nucleated cell, already fascinating in its mixture of living beings. But plant, animal, and fungal life greatly expanded the complexity of the free-living protist cell by repeating it to make multicellular copies that ultimately evolved into separate tissues, such as reproductive and nerve tissue, with distinct functions.

The offspring of some of these protists, starting in earnest perhaps a thousand million years ago, failed to separate after they reproduced by cell division. They began permuting themselves into colonies, some of whose members died each generation. Thus certain colonial protists became physically large members of the group, and the diversity of protoctists evolved. Looking at modern protoctists suggests how such colonies could have formed from individual cells. Animals, including of course ourselves, are transformed colonies of protist cells.

Charles Darwin emphasized that evolution occurs as different individuals pass on their traits by out-reproducing others. But individuality, always in flux, is relative. Cells form and interact in a wide array of configurations. Together they form individuals at various size levels and degrees of interdependence. The alga *Chlamydomonas*, with its large green single chloroplast, is a bacterial composite. *Volvox*, a spherical confederacy of *Chlamydomonas*-like protist cells, is a green multicellular descendant of *Chlamydomonas*, just as animals are multicellular descendants of swimming protists. [PLATE 40]

The origin of any "individual" large organic being depends on integrative gene-transferring processes not easily reversed. These integrative processes first stabilized as the colonial protoctists evolved from free-living protists. *Volvox* algae, like other protoctists, fungi, plants, and animals (but unlike bacteria), do not casually trade their genes. Larger organisms simply cannot trade genes the way bacteria do.

Any single protoctist, plant, fungus, or animal is a member of a species. Most likely, protoctists were the first organic beings to form species and the first whose species went extinct. The origin of individuals who all belong to the same species is identical to the origin of the first protoctists. Canadian microbiologist Sorin Sonea makes a good point when he claims that bacteria, because on a planetary scale they reversibly trade genes, do not have true species. Species are groups whose members interbreed. All bacteria on the planet can in principle interbreed. If anything, they might be said to form a single, global species.

Species demarcation is thus much more applicable to the protoctists, in which, indeed, it first

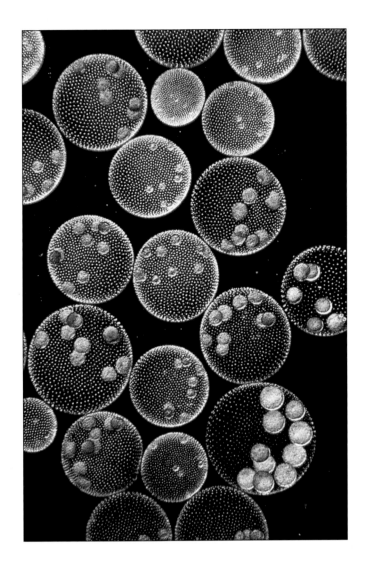

40
Colonies of *Volvox*.
Phylum: Chlorophyta
(green algae). Kingdom:
Protoctista. Individual cells
of this green colonial alga
resemble free-living cells
of *Chlamydomonas*. The
evolutionary move from
unicellular to multicellular
"individuality" is a crucial
one that has occurred
many times. It may be happening again as electro-
nically communicating,
technologically interacting
human beings form net-
works required for survival.
(See chapter 9.)

41

Naegleria, a protist. Phylum: Zoomastigina. Kingdom: Protoctista. Scanning electron micrograph of the *Naegleria* amoeba caught in the act of attempting to eat its conspecific neighbor. In evolution, ingestion followed not by digestion, but continued internal existence of the devoured being was an important means of starting up cellular symbiosis. When protists of the same species devoured but did not digest each other they sometimes merged nuclei and chromosomes, a comingling equivalent to the first act of fertilization or mating.

appeared; so did sexuality—of the "meiotic" kind. Fatefully for the future history of life forms such as ourselves, in protoctists sexuality became inextricably linked to death. Bacteria can be killed but they do not naturally die. Certain protoctists, notably ciliates and slime molds, unlike bacteria, will age even if external conditions are suitable for health. Aging and death, in which living cells disintegrate with predictable timing, first evolved in sexual protoctists. "Programmed" death as the final stop of a lifelong metabolism was absent at the origin of life—and for a very long time afterward.

Unlike us, bacteria are immortal; they will live until external conditions prevent autopoiesis. By contrast, like us, many protoctists age and die at the end of a regular interval. Aging and dying is an internal process, thanosis, and it arose in our microbial ancestors at some time during the evolution of sexual individuals. Strange to say, death itself evolved. Indeed, it was the first—and is still the most serious—sexually transmitted "disease."

Sexual Genesis in the Microworld, or When Eating was Sex

Animal sex always involves meiosis. In meiosis, as in mitosis, chromosomes attach to spindle microtubules and are partitioned out to offspring cells. However, meiosis omits a crucial step: the doubling of the chromosomes.

Meiotic cell division thus produces two offspring cells each with only half the number of chromosomes that were present in the original parent cell.

For example, after meiosis a human cell with the standard allotment of forty-six chromosomes becomes an ovum or sperm cell with only twenty-three, ready to find its "other half." Meiosis, which halves the number of chromosomes per cell, and fertilization (after sexual intercourse, flower pollination, protoctist or fungal conjugation), which doubles the number, must go hand in hand.

Meiotic sex evolved in mitotic protoctists long before any animal appeared in the record of life. Nevertheless, some modern protoctists show neither mitosis nor meiosis. The giant freshwater ameba, *Pelomyxa palustris,* for example, reproduces by pinching in half its many nuclei as it pinches in half its body—a single huge cell. Dinomastigotes ("dinoflagellates,") undergo a unique variation on the theme of mitosis: their DNA is not coated in histone protein and their chromosomes, unlike those of other microbes, remain visible throughout a cell division in which the nuclear membrane remains intact.

Meiosis is a variation on the theme of mitosis. Meiosis likely evolved in doubled cells that already divided by mitosis. The first fertilization event probably satisfied an urge not to merge but to eat. This could have happened if protist cannibals ate one another. Microscopists sometimes witness microbial wranglings in which a hungry cell engulfs a neighbor. [PLATE 41] But the cells do not always digest what they engulf.

Harvard University biologist Lemuel Roscoe Cleveland (1898-1971) saw cannibalized protists live on although half-devoured. The protists he studied, covered by 9(2)+2 undulipodia, are called hyper-mastigotes. Normal hypermastigotes, which contain only a single set of chromosomes, live in the swollen hindguts of wood-eating termites and cockroaches. Cleveland saw the hypermastigotes engulf one another. He noted that once they did, their mem-branes merged to make them into doubled cells. Most of the doubles died. But Cleveland also saw that some of these doubled beings reproduced. Although sloppily, a doubled microbial monster would undergo cell division and give rise to another doubled microbial monster.

Cleveland saw how thwarted cannibalism could have led to the first sets of doubled chromosomes. Moreover, abnormal cell divisions—a precursor to the meiosis that takes place in our own cells—could restore the merged would-be cannibals to their orig-inal single-set of chromosomes. Such potential steps in the evolution of sex occurred in a few hours in the protists of Cleveland's laboratory. In nature, unseen and long ago, they must have occurred spon-taneously many more times. Certain hypermastig-otes today when they fuse form hard-walled resistant structures, cysts, capable of withstanding deprivation. This double form—perhaps originally deriving from cannibalism—would have protected the proto-sexual beings engaging in it.

Once upon a time, we think, eating and mating were the same. Terminal microbial indigestion may seem rather unromantic as the source of the human sex drive. But Cleveland's picture of hungry, serendipitously mating hypermastigotes presents a mix of comedy and terror appropriate for the origin of sex. In times of scarcity, our unicellular Ur-parents would have desperately fed on one another. Sometimes their membranes would have merged. Doubled, partly doubled, and huddled in their walled cysts, they waited. Abnormal beings, some now with faulty sets of chromosomes, many would die. Those returning to the original one-set state

would be naturally selected; only they could begin to reproduce normally again. The doubled and other monsters would tend to die. All the same, abortive cannibalism and other fusions would often afford protection against seasonally scanty food and periods of desiccation. Unable to enter the sturdy quiescence of a sex-induced cyst phase, those reluctant to eat their kin starved and died of thirst.

The cells in our own animal bodies are in a diploid, or double chromosome, state except for the protist-like ova and sperm, which are in a haploid, single-chromosome state. Each animal body is a sort of diploid husk, morbidly discarded by those haploid sex cells that manage to produce each generation a fresh new body and thus continue beyond the death of the "individual." The diploid body pays the ultimate price—death—for transmission of haploid sex cells.

Perhaps originally cannibals in distress, chromo-somally doubled protists are our ancestors. Humans and all animals inherited death from these early eukaryotes. Each generation starts where the last left off and, depending on who survives, each follows a slightly different course. Over time, this leads to new species.

Aggregated into colonies, layers of eukaryotic cells eventually evolved into tissues. It is an amazing fact that all zygotes—fertilized egg cells that survive to become embryos—grow into plants or animals composed not just of many cells but of several (or many) kinds of cells that assemble into distinct tissues. Although far more impressive in animals and plants, protoctists too may be composed of several sorts of cells showing division of labor. Protoctists recognize one another; when deprived of water or food, member cells of the same species can congregate. Take slime molds: these bacteria-eating amebas exchange metabolic information, seek out their own kind.

When food is aplenty, the amebas feed alone. But when food is depleted, each hungry cell secretes a compound that is mutually attractive. Amebas move in the direction of highest concentration. Together the cells merge to form a moving "shmoo" that grows upward in a slimy mass until its "head" bursts and releases sturdy-walled cysts, safely transportable by wind and water. A new generation of amebas will begin if the cysts land in a clement environment.

The Power of Slime

A focus on our protoctistan origins is humbling. There is no denying our kinship to this sort of living matter. Humans are integrated colonies of ameboid beings, just as ameboid beings—protoctists—are integrated colonies of bacteria. Like it not, we come from slime.

An estimated quarter of a million protoctist species dwell in lakes, rivers, water falls, warm springs, damp soils, transient puddles, dew, and frost, and on the walls of swimming pools and pipes. Plants, animals, and fungi—the visible kingdoms—began as mere fiefdoms of the Kingdom Protoctista. The initial members of this kingdom had already evolved a thousand million years before spinning off the remaining three kingdoms of life.

Today crystalline beauties such as diatoms, foraminifera, and radiolaria inhabit Earth's oceans. Because protoctists live most abundantly in the tropics, surely far more species are unknown than known. The best studied include a few notorious killers, such as the trypanosomes that cause sleeping sickness, Chagas' disease, or leishmaniasis. In the latter, a disfiguring disease of the tropics, growing, swimming protists may devour embryonic mucous membranes *in utero*, causing human babies to be born without mouth or nose. But together most protoctists cradle the planet in their gentle biochemical grip. They carpet the ocean with food and supply its oxygen, churn soil, clear surfaces of bacteria. They cycle sulfur, phosphorus, silica, and carbon on a global scale.

Protoctists partake of planetary physiology. Perhaps the most numerous are the coccolithophorids. Though microscopic, these single-celled algae are one of the few forms of life distinctly visible by satellite. A "bloom" of coccolithophorids may whiten the green waters off the coast of Europe for two hundred kilometers. [see PLATE 13]

Not until the ocean water is concentrated by centrifuge in the laboratory and the sedimented sample magnified some ten thousand times does the source of the white patch reveal itself: coccolithophorid scales. Each coccolithophorid is patterned with hundreds of chalky "buttons." The scales, and the spaces between them, may be the natural equivalent of Venetian blinds, serving to provide the alga's plastids with an optimal quantity of sunlight. The scales released by dying microbes—millions of buttons per milliliter—change the water to a milky color seen better by satellite than by ship. [see PLATE 14]

Because salt accumulates in its cell and can destroy it, a coccolithophorid must make complex sulfur compounds that balance internal ion concentrations. These sulfur compounds are unstable; they break down to form dimethyl sulfide, a waste gas wafted into the air. Once released, dimethyl sulfide gas reacts with oxygen to produce tiny aerosol particles of sulfate. These particles seem to be involved in the formation of cloud cover by serving as nuclei for the condensation of water vapor. Since the cloud cover radiates back to space and leads to cooler temperatures, a bloom of coccolithophorids such as *Emiliania huxleyi* may act as a global air conditioner.

Vast amounts of materials flow through the bodies of coccolithophorids and other photosynthesizing protoctists. Protoctists, not plants, serve as the base of the entire marine food chain. Floating protoctists attend to the needs of marine ecosystems far offshore. Substrate-dependent species support the bulk of communities near shore.

Building their skeletons of chalk, glass, organic

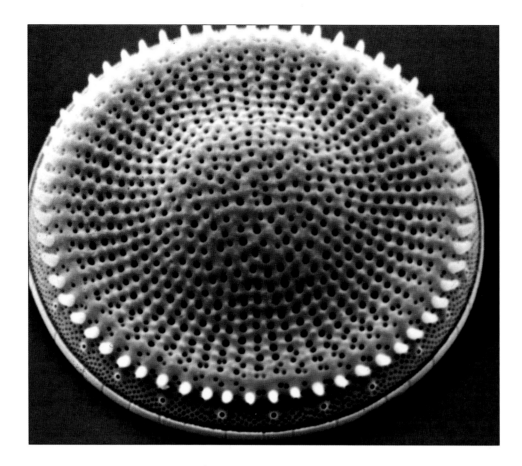

42

Stephanodiscus, a diatom.
Phylum: Bacillariophyta.
Kingdom: Protoctista.
Radially symmetrical pillbox diatom. Diatoms, usu-
ally tan or brown in color
and prevalent in oceans,
deplete water of silica as
they make their gorgeous
microshells.

fiber, and even exotic salts like strontium or barium
sulfate, some protoctists mine the oceans for trace
chemicals. They transform seascapes when, after
building up their bodies' hard parts, they die in
droves. Diatoms deplete silica from the ocean on a
global scale to create their exquisite pillbox forms.
[PLATE 42] Radiolaria form opaline shells that fall like
rain through the water column, making a film that
later hardens into a flintlike sedimentary rock called
radiolarite. Foraminifera make up part of the
limestone rock composing the great pyramids
of Egypt.

Protoctists are cosmopolites, infiltrating
themselves into the great water and soil mass of
Earth. Like the Sphinx, protoctists are recombined,
fused beings. The first organic beings to form
sexually reproducing species, their cellular quirks
are at the core of human sexuality. The protoctists
bequeathed to all subsequent kingdoms of life the
physiological necessity of death. They, along with
bacteria, are the supreme architects of the living
global environment.

So, what is life?

Life is the strange new fruit of individuals evolved by symbiosis. Swimming, conjugating, bargaining, and dominating, bacteria living in intimate associations during the Proterozoic gave rise to myriad chimeras, mixed beings, of which we represent a tiny fraction of an expanding progeny. Through corporeal mergers disparate beings invented meiotic sex, programmed death, and complex multicellularity.

Life is an extension of being into the next generation, the next species. It is the ingenuity to make the most of contingency—to make animals, for example, out of a botched attempt at cannibalism. Life is bigger than the cell, the organism. It includes the biosphere, the surface planetary environment as a whole—from the formation of marine clouds to the control of ocean chemistry by protoctists and their progenitors. ▪

6

The Amazing Animals

Can we believe that natural selection could produce,
on one hand, organs of trifling importance,
such as the tail of a giraffe, which serves as a fly-
flapper, and, on the other, organs of such wonderful
structure, as the eye, of which we hardly as yet fully
understand the inimitable perfection.

— **Charles Darwin**

Full fathom five thy father lies;
Of his bones are coral made:
Those are pearls that were his eye
Nothing of him that doth fade
But doth suffer a sea-change
Into something rich and strange.

— **William Shakespeare**

And, striving to be Man, the worm
Mounts through all the spires of form.

— **Ralph Waldo Emerson**

The (Bower) Birds and the (Honey) Bees

Animals are multicellular assemblages that always develop from one single cell. A swimming (undulipodiated) sperm penetrates and fertilizes an egg. The fertilized egg then divides to form two, four, eight, and more cells that make a blastula embryo, the essence of animality. [PLATE 43, *page 120*]

Each animal phylum (such as the jointed-foot arthropods which include insects, spiders, crayfish, and crabs) is a great group—or a formerly great group. The brachiopods, for example, are clamlike but nonmolluscan creatures often found as fossils in Paleozoic marine rocks. Our own phylum, the chordates, includes sea lemons and jawless fish as well as salamanders and pigeons. According to one scheme, some thirty-three phyla exist with animals that are alive today; others have members that have long been extinct.

We have explored the origins of life, the bacterial realm, the protoctists. Moreover, our discussion has been chronological, following evolution and the rise in complexity. So, why now—well before the end of the book—do we talk of animals?

Fossils reveal that animals evolved earlier than plants or fungi. Animals—exclusively marine animals—began leaving a rich fossil record in the early Paleozoic. But there is no trace of plants or fungi until more than three hundred million years after shelly animals appeared. Even today, animals—typical in this regard, since life evolved in water—abound more in water than on land. Only plants and fungi are paradigmatically creatures of the land. Microbes had to venture onto land before these newer kingdoms could evolve there.

An animal cell is simple, relative to that of any plant. Plants, animals, and fungi all sequester their genes into nuclei; all have mitochondria organelles to handle oxygen respiration. But plant cells have the additional complexity of an organelle for tapping the energy of the sun. Animals lack such plastids. Nevertheless, members of Kingdom Animalia boast a proud heritage. Indeed, sensing and acting, they have evolved some amazing capabilities.

Insectivorous bats detect their prey in the night skies by emitting sounds at frequencies too high for humans to hear. They gauge, from the echoes that come back to them, the presence and position of objects in their surround. "Blind" when their ears are plugged, their ultrasonic picture of the world may resemble our ultrasound scans of unborn babies. Bat-blindness was discovered by the same soup-boiling Lazzaro Spallanzani we encountered in chapter 3.

Not above incorporating colored poker chips into their constructions, the male bowerbirds of Australia and New Guinea build elaborate colored mounds. Wings flapping and prancing, the bowerbird calls loudly, fluffing up his feathers, flaunting about. To assure a female chooses him, he spruces up his ten-foot-square, foot-high bower with bark, crushed fruit, charcoal, scavenged feathers, or even blue laundry powder. To entice her, he decorates the bower with bits of snail shell or fresh flowers he replaces daily. Nonetheless, when she ventures in, he mates with her so violently it damages his own bower, and then, when finished, he pecks, claws, and expels her from his bachelor pad. This tragicomedy of lust is completed as the male embarks on a proprietary strut about his territory, taking time out only to destroy the bowers of younger males.

Some crabs acquire stinging sea anemones and wield them as weapons against potential predators. Herring gulls drop clams, whelks, and other shellfish from a height of several meters onto rock to crack open their hard shells. Japanese green-backed herons lure minnows by dropping twigs on pondwater. Dogs sniff, bark, and scamper in their sleep as if chasing rabbits. Anesthetized and painted with an obvious spot on their forehead while unconscious, chimpanzees and orangutans (but, strangely, not gorillas) have awakened to wipe their brow looking in a mirror—proof positive that they recognize

themselves. Subordinate simians hide their copulations from dominant males. This suggests that they not only recognize themselves but have a mental model of how others see them. Vervet monkeys sound three distinct alarm calls, depending on whether the object of their concern is a leopard, a python, or an eagle. Dolphins whistle at an amplitude and frequency that is distinctive for each individual; they seem to call each other by name. Captive in an aquarium, one bottle-nosed dolphin has been reported to imitate, perhaps for comic effect, the swimming postures of turtles and penguins. This dolphin, who used a gull feather to scrape algae on an underwater window, apparently mimicked a human diver. The dolphin even emitted a stream of bubbles like those the diver expelled.

But exceptional brain power is not always at the root of exceptional communicative and apparently intelligent animal behavior. Honeybees see in color (including ultraviolet, which humans can't see— but not red, which we can). They orient themselves by light polarization in the sky. Cooling the hive by regurgitating water droplets and fanning their wings, and warming it by shivering, they are masters of environmental regulation. Honeybees seek pollen grains and nectar and the workers that discover rich sources of these materials communicate their finds to others. The successful scouts appear to perform two distinct dances to specify the location: one circular, for close food sources, the other a "waggle" dance, for food more than about a hundred meters away. The assumption that this instinctive bee behavior, the discovery of which earned Austrian-German zoologist Karl von Frisch a Nobel Prize in 1973, is no more conscious than a computer program, has never been proven. Honeybee unlearned knowledge of dances seems to be innate, but it is nevertheless possible that the bees are in fact "aware" of their significance.

Animals are so amazing that we need not, as humans, consider ourselves as anything more than

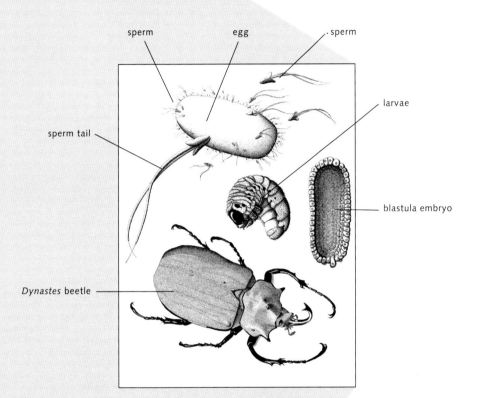

sperm

egg

. sperm

sperm tail

larvae

blastula embryo

Dynastes beetle

43
Dynastes, a beetle.
Phylum: Arthropoda.
Kingdom: Animalia. The
hairy segmented structure
is the larva, or grub,
formed from the hollow
mass of cells known as the
blastula embryo shown at
right. The blastula, a
defining characteristic of
all animals develops from
the sperm-fertilized egg,
which undergoes many cell
divisions.

animals in order to feel justifiably proud. But as Donald Griffin laments, ever since Darwin forced humankind to recognize its animal kinship, the inclination has been just the opposite:

> Much of twentieth-century science has gradually slipped into an attitude that belittles nonhuman animals. Subtle but effective nonverbal signals to this effect emanate from much of the scientific literature. Physical and chemical science is assumed to be more fundamental, more rigorous, and more significant than zoology. Modern biology revels in being largely molecular, and this inevitability diverts attention away from the investigation of animals for their own sakes. Part of this trend may be due to an unrecognized reaction against the deflation of human vanity by the Darwinian revolution. The acceptance of biological evolution and the genetic relationship of our species to others was a shattering blow to the human ego, from which we may not have fully recovered, for it is not easy to give up a deep-seated faith that our kind is unique and qualitatively superior.

Griffin goes on to suggest:

> A psychological palliative that may be subconsciously attractive, even to many scientists, is to shift attention away from the embarrassing fact of our animal ancestry by accentuating those aspects of science that are more akin to physics. This may help explain why so many appear to be so certain that consciousness and language are uniquely human capabilities and that the discovery of symbolic communication by honeybees "upsets the very foundation of behavior, and biology in general." Quite the contrary, such discoveries in the field of cognitive ethology extend and improve our understanding of animals; a definition of biology that rules out those discoveries a priori suffers from self-inflicted impoverishment.[1]

Consciousness, a private affair, is not directly measurable. But an inability to render a quality measurable is no reason to assume its absence—to assume that animals are mere instinct machines. Indeed, we would take Griffin a step further. Not just animals are conscious, but every organic being, every autopoietic cell is conscious. In the simplest sense consciousness is an awareness of the outside world. And this world need not be the world outside one's mammalian fur. It may also be the world outside one's cell membrane. Certainly some level of awareness, of responsiveness owing to that awareness, is implied in all autopoietic systems. The world, after all, is not a petri dish; the sky does not rain agar. To live every organic being must sense and respond to its surrounds.

All animals have a multicellular, multitissued stage in their life histories. But the complex chemistry of life is still confined to the cell, minimally less than a single micron in diameter. Each generation of every kind of animal returns to the ancestral-type single cell as fertilized egg. The increase in size and complexity of animal life has occurred as socially aggregated cells evolved to form unified bodies.

Animals eat other organisms. They do not enjoy the vegetal luxury of feeding themselves through photosynthesis but must go out into the world—begging, borrowing, and stealing—in order to accomplish what plants can do sitting calmly in one place. Many extraordinary animal attributes stem from the requirement to procure hard-to-get food. Others may be traced to the reproductive imperative of sexuality: in each generation sperm and egg must join. The aesthetics of bowerbirds, the dancing of honeybees, the dreaming of dogs have all evolved. Connected to the procuring of food and/or mates, all such behaviors manifest the ancient autopoiesis of self-perpetuating life.

Animal behaviors seem directed, "purposeful." Yet like bacterial magnetotaxis and protist cannibalism, they may also be perceived as ways in which the

44

Lima scabra, a scallop. Phylum: Mollusca. Kingdom: Animalia. A picture of an adult showing the soft parts of the animal. This bivalve mollusk develops from a planktonic larva (a ciliated trochophore larva) that develops in turn from a blastula embryo. A blastula embryo is a defining characteristic of all true animal life. In spite of our experience on land most animal phyla have ocean-going members such as this one that dwells off the shore of Puerto Rico. The first, soft-bodied animals are thought to have evolved in the oceans over 600 million years ago.

universe's dissipative systems make waste of useful energy by putting it in the service of localized pockets of stunning order. Communication, imitative learning, tool use, and conscious thought make sense from a thermodynamic perspective.

What Is an Animal?

Most species of living beings today are probably animals. Estimates range from about three to more than thirty million species in this one kingdom. Some phyla, conspicuously successful, are well known: the echinoderms (starfish, sea urchins, sea cucumbers), the mollusks (clams, snails, squid), the chordates (fish, reptiles, birds), the coelenterates (hydras, corals, jellyfish), the arthropods (insects, spiders, crabs). Lesser-known phyla include the pogonophorans (some of the deep sea worms), onychophorans (velvet worms on the South American forest floor), and pentastomes (which live in the nasal passageways of mammals).

Animals—all with two-parent sex and fertilized eggs that form embryos—mature into reproductives subject to individual programmed death. [see PLATE 43.] Despite their diversity and exuberance, animals are evolutionary newcomers. The first animals evolved in an oxygen-rich world with large continental masses and open seas—a world not too different from that which supports us today. But at the time of their appearance eighty percent of the story of life—thus far—had already been played out.

Ocean dwellers from the start, and mostly ocean dwellers today, no animals appear in the fossil record until the end of the Proterozoic eon, some 600 million years ago. The most famous early animals, the Cambrian sea creatures, are even more recent; their fossils date to a little more than 500 million years ago. Only a very few ancestors of modern animals—some chordates, worms, arthropods, and mollusks—succeeded in leaving the ocean behind to survive on land. [PLATE 44]

Land animals—with their intricate bodies, devious minds, and sometimes elaborate societies—seem to have evolved furthest from the earliest cell. But consider: an animal is the very creature telling this story of evolution. Might that teller be just a tad bit biased in favor of its own kingdom? Perhaps, considering the source, the notion of progress from "low" bacterium to "high" human is a delusion of grandeur. As paleontologist Stephen Jay Gould once remarked, an intelligent octopus would probably regard eight arms as more perfect than two.

The evolution of the first animals is a fascinating question. But what is an animal? How might we have recognized the first member of the new animal kingdom in its own day? Certainly an animal is not just an organism that moves but does not photosynthesize, because most bacteria and many protoctists would then, by definition, be animals. What makes any animal distinctively and uniquely an animal? What has a person in common with a worm, a starfish, and a million kinds of beetle?

Animals, whether in the urban dark of a barroom or by a moonlit equatorial reef, all share the same life cycle. Fusion of the two different-sized cells, egg and sperm, begins the process of animality. The sperm and egg merge into a single fertilized egg that divides by mitosis to form a ball of cells. As cell division continues, the fertile egg becomes a hollow ball of cells — the blastula embryo. [PLATE 45] An embryo distinguishes the animal and plant kingdoms from the other three kingdoms; the blastula embryo distinguishes the animal from the plant. A plant embryo is a solid mass within the maternal tissue; an animal, blastula embryo is a hollow ball of cells usually existing free of mother-tissue in the water.

Animal embryos assure the animal kind of individuality. Even sexual protoctists do not require sex to reproduce; moreover, they never make the sexual product we call an embryo. Embryonic development is not part of protoctist individuality. Loose colonies of protoctists tend not to have fixed body sizes and shapes; rather, pieces of protoctists break off and form new amorphous wholes.

45
Fruit fly (*Drosophila melanogaster*) embryo prepared so that a red dye tracks the presence of the nervous system in the young animal. Because they can be grown (in pint milk bottles in the laboratory) from egg to adult to egg again in just a few weeks, more is known about the genes and chromosomes, the growth of the nervous system, the muscles and hormones, the sense organs, the mating behavior, and all other aspects of the biology of this insect than for any other animal, including humans. Phylum: Arthropoda. Kingdom: Animalia. The fruit fly is part of the varied and successful arthropod phylum, a group which includes not only all insects and spiders but also crustaceans such as crabs and lobsters.

Composed of tissues with distinct connections between the cells—desmosomes, gap junctions, septate junctions, and the like—the bodies of animals are individualized. Such cell-to-cell connections, unknown in any other kingdom, must be produced by embryo development. In a predictable sequence, cells of the embryonic animal divide by mitosis, they roll over one another and set up alliances such that many, sometimes most, of the animal's body cells must die in a preprogrammed fashion. If the young embryo cells do not die on cue or do not establish definite connections to send specific signals through their junctions, no animal body develops. Animal embryos are crucial.

In animals the muscles, nerves, and circulating fluids such as blood all derive by mitotic cell division and differentiation from the blastula. Doomed to mortality, the blood, muscle, and nerve cells stop dividing; their programmed death contributes to the construction of the body and the continuing of the theoretically immortal egg and sperm. Without the blastula embryo formed from the fertile union of sperm and egg, animals would not exist. The blastula embryo is the animal universal in all thirty-three or so phyla. In the usual course of animal development the cells of the blastula continue to divide, move, and die. They form the next stage, the gastrula, which makes a new mouth at the front end of a digestive tube. The mouth is succeeded by a swelling, a stomach, and an anus in this unfolding process, called, logically enough, "gastrulation."

The vast majority of blastulas in the thirty million animal species "gastrulate" and end up with a distinctive "tube-within-a tube" digestive system, committed to heterotrophic nutrition. Formed by cell-to-cell communication via gap or septate junctions with nerve cell synapses, the tube or gut establishes the ingestive nutrition so nearly universal in the kingdom. But a few kinds of exceptional animals lack intestines and even these must pass through a blastula stage. This kind

of embryo is the hallmark of any animal.

Neither fungi nor protoctist, although most are heterotrophs, forms any embryo at all. Probably correlated with the absence of the embryo is the absence of animal-style distinctive individuality. Plant individuality is far less fixed than that of animals. Although all members of the plant kingdom do form embryos, plant embryos are very different from any blastula. Each plant cell is walled from its neighbor, precluding the movements and realignments that all blastulas undertake as they become individual animal larvae or adults. An embryonic plant cell cannot form gap or septate junctions, nerve synapses, or any other of the animal cell-to-cell connections. Stationary in its place, the plant cell only grows, dividing by mitosis, or dies. It is the fateful blastula that presages all the nuances of animal behavior and distinguishes our kingdom from all the rest of life.

Great Grandparent *Trichoplax*

To the bacterial realm goes the award for metabolic innovation. As keepers of the biosphere, prokaryotes are the most inventive forms of life, and their descendants include the now-essential organelles within our own cells. The protoctists, too, dealt originally with the problem of environmental threat: in autopoietically changing to stay the same, they arrived at death, sex, and metamorphosis into resistant structures. But with the origin of animals nature seems to have reached new levels of playfulness, awareness, complexity of form, responsiveness, and deception.

A butterfly's wing bearing an imitation raindrop with a line displaced just as if it had been refracted through real water; a cheetah poised to pounce; an acrobat juggling upon a high wire: animals amaze.

Today's minimal animal is *Trichoplax*—a headless, tailless creature discovered crawling on its belly along the side of a marine aquarium in Philadelphia in 1965. Were it not for its sex life and its embryo,

1

Trichoplax would be a protoctist. Wafted along by its undulipodia, *Trichoplax* superficially resembles a slime mold slug or a giant ameba. But it is a multicellular being throughout its life and is a true animal. It has more undulipodia on its belly than its back. (Here it resembles an insect, which has more undulipodia on its legs than its antennae.) Having neither head nor hind end, right nor left side, no eyes, no stomach, this minuscule slow crawler gives away the secret of its animality only at reproduction. After fusion with sperm, a spherical *Trichoplax* egg becomes a blastula embryo that, with more cell division, flattens and, ameba-like, slinks away. Although you probably wouldn't want to hang a portrait of it in the drawing room, *Trichoplax* likely bears a strong resemblance to our earliest animal ancestors.

A sponge is an animal composed of only a few types of functionally and morphologically different cells. For example, those on the outside may grow glassy rods for support and protection; those on the inside use their undulipodia to maintain a flow of water from which food can be extracted. If yellow and orange (*Haliciona*) sponges are squeezed through cheesecloth so that each is broken into bits and mixed with the other, the cells manage to find their own kind in the aqueous environment. After a few hours the cells reorganize into fully formed and distinct yellow and orange sponges. So too, a freshwater polyp, a jellyfish relative having about a hundred thousand cells of a dozen cell types, can be disassociated into single cells. In permissive solutions they begin to rearrange themselves. Unlike sponges, they cannot complete the process. Monstrous growths, in which head, gut, and foot (basal stalk) realign unsuccessfully, are the result. In this case, the integrating mechanisms assuring autopoietic self-maintenance fail.

In most colonial green algae and ciliates (all of which are protoctists) any single cell may separate and reproduce on its own. In others, only certain cells reproduce. The theme of animal evolution, the development of discrete individuals, involves curtailing reproduction in favor of specialization. Protoctist anarchies, in which any cell could reproduce, were replaced during the emergence of animals by cell oligarchies, in which only a few (sometimes a very few) had the privilege of living on into the next generation by way of progeny.

The transition from cell, to cell society, to animal organism is an old story in evolution: individuals group into societies, which themselves become individuals. Under intense selection pressures, swimming protists became colonial protoctists. Then, in the later Proterozoic eon, *Trichoplax*-like animal bodies appeared. The specialization of massive numbers of cells into integrated individuals is at the base of animal life—and of those later groups, fungi and plants.

Sex and Death

Only accidental, externally caused death, existed at the origin of life. So it was for a long time thereafter. But with protoctists came "programmed death": death in which cells age and die as part of the life of the individual. In familiar animals—insects, mammals, and birds—the difference between the part that dies and the part that potentially lives on is the difference between the body and the sex cells. In mammals the sex cells (or "germ plasm," as biologists sometimes say) are the only cells whose direct progeny survive into the next generation. In contrast to the ova and sperm, the "soma"—the animal body—has a discrete life span.

With a high degree of precision, animal cells must reproduce—or cease reproducing. For example, during the intrauterine development of the mammal brain more than ninety percent of the cells that develop die before the fetus becomes an infant. These brain cells stop growing and disintegrate, are sacrificed in the process of growing a healthy infant. The essential difference between the living germ

cells and the dying body cells of animals is likely
very old.

We speculate that the ancestors of animals were
composed of relatively few cells that differentiated
into at least two distinct kinds. One kind specialized
in using their 9(2)+2 microtubule organelles to form
undulipodia for propulsion, for sensing prey, for
fostering water flow over or through the animal, or
for sweeping food particles into and along digestive
systems. [PLATE 46] But it is an oddity of physiology
that once animal cells dedicate their centrioles to
forming undulipodial shafts they can no longer use
them to create the motility apparatus for mitotic cell
division. This means that animal cells stood to gain
by sticking together in specialized colonies. Even
today an animal cell, whether of a tissue or a sperm
after growing the undulipodium, no longer divides.
A centriole forms a kinetosome and relinquishes
cell immortality; a kinetosome cannot revert to a
centriole. The irreversibility of kinetosome
formation appears to be an inviolable rule within
the animal kingdom. Animal cells can either form
kinetosomes (grow undulipodia from centrioles) or
reproduce by mitosis — but not both. An animal
cell with a kinetosome is a dead animal cell — its
days are numbered, as it will not divide again.

Perhaps the DNA reported by David Luck and
John Hall to be in the kinetosome-centriole is used
for mitosis or to form an undulipodium but not
for both. Like choking from inhaling water, any
attempts of cells to simultaneously reproduce and
maintain undulipodia would have been thwarted.
And yet animals seem to have found an answer to
this genetic dilemma: by sticking together in
colonies — colonies where some cells reproduce
while others form undulipodia — they could in effect
have their cake and eat it too. The restriction of a
cell unable to divide after growth of its 9(2)+2
organelle was overcome by colony formation.
The great majority of cells retained their option to
divide, while a few sacrificed immortality to be

46
Undulipodium in cross
section. The shaft
(axoneme) displays the
9(2)+2 arrangement of
microtubules. This distinct
intracellular organization
is found in widely diverse
beings throughout the
natural world from the
sperm cells of men to
those of ginkgo trees.
Electron micrographs of
cuts through the shafts
of the cilia propelling
swimming paramecia and
trichomonads and the cilia
that push the egg through
a woman's Fallopian tube
also reveal this 9(2)+2
pattern.

undulipodiated. But even the cells that do divide in the animal do not do so indefinitely. After 600 million years the adult animal is still a mated protist's way of making other mating protists.

Our whole life from womb to tomb is in fact an interim stage in the life cycle of tiny fused cells. Animals emerge into another dimension, visible life and consciousness, only to return via sex to their ancient single-celled, microbial state. Death is the price we all pay for this ancient history of multicellular compounding, for this inability of hungry protists to undo their Proterozoic entanglements. What "dies" is the body, the adult flesh after it has released into the water or body fluid the protist-like tailed sperm and chubbier egg. Animal life did not appear de novo, but from protoctist predecessors. Protoctists with elaborate cycles of fertilization, multicellularity, and meiosis became animals.

Like programmed death, gender is not intrinsic to life. Gender evolved. Cells of different mating types, like protoctist lovers today, were initially identical to each other in appearance. The seasonal merging and restoration of chromosomal numbers in fertilization set the stage for the origin of gender. The first mates met slapdash in a watery environment then, as protoctists do today. Responding to slight chemical differences in each other, mates came together. Sponge, sea urchin, fish, and even mammalian sex cells, like their protoctist ancestors, still meet in watery places.

Animal cells continue their ancient practice of aquatic encounter. The sex cells of oysters and even some frogs and fish meet directly in the water, to fuse unattended by adult bodies. In reptiles, birds, and mammals, however, sexual mergers occur *in vivo*. Genitals evolved independently in many animal lineages. The penis or intromittent organ of the male created a delivery system for sperm. The female genital tract afforded the ova a protected place where fusion could occur. The many small sperm of males compared to the few larger eggs of females was the beginning of an evolutionary

asymmetry which today expands into the realms of political, sociolinguistic, and psychological debate. Evolutionary biologists suggest that early sexual inequality—males maximize reproduction by inseminating the largest possible number of females, whereas after a certain limit mating becomes superfluous to females constrained by devotion to their lesser quantity of eggs—is behind distinct male and female attitudes toward sex.

Cambrian Chauvinism

English geologist Adam Sedgwick (1785-1873) named the time period to which the oldest fossils belonged Cambrian, after "Cambria," the old name for Wales in southwestern Great Britain. To him and other early paleontologists the appearance upon Earth of animals seemed miraculously sudden. All prehistory prior to Sedgwick's Cambrian became known as the "preCambrian." Until the late twentieth century the origin of the Cambrian animal fossils was considered "the most vexing riddle in paleontology."[2] So quick was the apparent appearance of animal life in the fossil record—not only in Wales but also in Newfoundland, Siberia, China, and the Grand Canyon of Arizona—that it is still referred to as the "Cambrian explosion."

Today, much of the answer to the riddle is known. Foraminifera and other conspicuous protoctists, the "protozoa" whose fossils were once dismissed as those of tiny invertebrate animals, in fact preceded animals by at least 500 million years. [PLATE 47] Because, like early animals, most protoctists were small and did not form hard parts, they remained largely undetected and unpreserved. Life prior to the Cambrian, despite its astounding biochemical and metabolic innovations, is still often dismissed as "preCambrian"—with the connotation that nothing much worth mentioning in evolution happened between the origin of life and the appearance of shelled animals.

Bacteria and protoctists set the stage. They, not

47
Giant acritarch, an unidentified fossil microbe, nearly a millimeter in size. This acritarch is found with the mineral phosphatite. It is suspected that the beings which fossilized as acritarchs were among the oldest protist cells. Animals did not appear *ex nihilo* but from colonial protoctists in soft-bodied organisms that left scarce traces in the fossil record. The Cambrian "explosion" had a long microbial fuse.

48
These fossils called sclerites appear at the beginning of the so-called Cambrian explosion. Whether they are fossil fragments of the very first animals or protoctist predecessors to the animals has not been established.

animals, introduced DNA recombination, locomotion, reproduction leading to exponential growth, photosynthesis, boil-proof spores. They, not animals, pioneered symbiosis and the organization of individuals from multicellular collectives. They invented intracellular motility (including mitosis), complex developmental cycles, meiosis, sexual fusion, individuality, and programmed death. The prokaryotic microbes, not animals or plants, still run all the geochemical cycles that make the planet habitable. The protoctists, in their new status as individuals from coevolved bacterial communities, invented resistant cysts, skeletons and shells, gender behaviors, cell-to-cell communication, lethal toxins, and many other processes later co-opted by animals. Animals were preceded by bacteria and protoctists, not by chemicals. The animal explosion had a long microbial fuse.

Patterned fossils of tiny shelly plates, known collectively as sclerites, mark the beginning of the Cambrian about 570 million years ago. [PLATE 48] The lowermost time-rock division or period of the Phanerozoic eon is the Cambrian. Distinctive animal fossils abound in Phanerozoic rocks overlaying barren Proterozoic strata. Sedimentary rocks from all over the world deposited 530 million years ago (still in the Cambrian) contain a striking array of skeletalized marine animals. At about this time, brachiopods (lampshells) and annelid worms appeared. So did trilobites and other arthropods (of which insects and lobsters are modern examples).

Some paleontologists still wonder how these various phyla could have cropped up "all of the sudden." Noting iron rust and other clues that oxygen had entered the atmosphere 2,000 million years ago, some scientists suggest that a threshold level of atmospheric oxygen (O_2) itself induced animal evolution. But any scenario to account for the "sudden" appearance of animals is almost surely a misreading of the evidence. Animals, although they evolved late in the history of life, did not evolve suddenly. "Seemingly," writes paleontologist Harry B. Whittington, "there was a long period of metazoan [animal] evolution before the Cambrian, but it is only in the earliest Cambrian rocks that minute shells of metazoans appear.... The Burgess Shale shows that it was not only metazoans with hard parts that were diversifying in the Cambrian, but also soft-bodied metazoans, including coelenterates, worms, arthropods, chordates, and various strange animals."[3]

The Burgess shale is a collection of Cambrian fossils exposed on a high mountain in Yoho National Park in British Columbia, Canada. Discovered by Charles Walcott in 1909, the exquisite and numerous Burgess shale fossils have given paleontologists lifetimes of work. Because even soft-bodied animals were preserved in it, this shale is a treasure. Shallow marine dwellers were preserved in the underwater mudslides that made the Burgess shale. These were a large variety of organisms, some similar to modern forms, others with no known descendants. Among the beautiful, if monstrous, animals are *Opabinia*, a five-eyed sea-bottom crawler with curved tail fins and a grasping, jointed organ that suggests it was a formidable predator, though only four inches long.

Hallucinogenia, in accordance with its name, has puzzled paleontologists because until recently no one was sure which side was up (spikes as armor) and which side was down (spikes as legs). Of the many Cambrian arthropods that the Burgess shales have preserved, only one sort gave rise to a lineage that much later evolved into the vast array of land creatures with six legs known today as insects. Had evolution taken another course, another Cambrian arthropod—or an entirely different animal, for that matter—might have gone on to populate the continents.

Among the most touching of the Burgess specimens is *Pikaia*, the first known member of our own chordate phylum—the phylum to which humans and all other animals with backbones belong. *Pikaia*, a segmented wormlike swimming creature, is inconspicuous compared to more spectacular Burgess forms. But it had a solid cartilaginous rod— the notochord—running down its back. This universal structure of chordates if not present in the adult is fleetingly apparent in larvae or other immature life cycle stages. Until the discovery of *Pikaia* in the Burgess shale, no chordates were known from any rock older than about 450 million years, deposited during the Ordovician, the geological period after the Cambrian.

The Burgess chordate is a stunning discovery because it shows that the predecessors of darting *Pikaia*—which may have been the ultimate ancestor of all fish, amphibians, reptiles, birds, mammals, and us—were alive and navigating the muddy waters 530 million years ago. The success of *Pikaia* may be directly responsible for the later emergence of such a wide diversity of forms, from mugwump to turtle, moose, rabbit, and giraffe. The presence of streamlined *Pikaia*, with its pointy, somewhat cobralike flattened head and bifurcating snail-shaped tail, shows that our ancestors swam in the primeval oceans.

Before the armor-headed trilobites crawled the planet, before droves of clam-like lampshells expired in Cambrian muds, before eurypterid "sea scorpions" left their hard exoskeletons in the fossil record, soft-bodied animals proliferated. Even less obvious and far older than the Burgess shale animals are the "Ediacaran" beings preserved in sandstones 700 million years old—before the Cambrian period, before the Phanerozoic eon. Most are probably not animals at all but bizarre, extinct protoctists. In the 1950s Martin Glaessner of the University of Adelaide named these extraordinary fossils after a rock formation in the Ediacara Hills of Australia. Similar soft-bodied beings have been found in England, Namibia, Greenland, Siberia, and some twenty other localities.

The Ediacaran organisms seem to have been floating gelatinous beings enjoying shallow water at sandy beaches. Some were flat, others "quilted," others intricately textured organisms. They ranged, shapewise, from leaflike *Pteridinium* to three-armed *Tribrachidium*. But these Ediacaran beings seem to have formed no hard parts, eggs, sperm, or blastula embryos. They may have been large protoctists, animals, or both. Some of the larger Ediacaran beings probably photosynthesized in shallow coastal seas. Others fed on bacterial pastures. But their lack of armor indicates that large predatory organisms had not yet evolved—it was truly a "Garden of Ediacara."[4]

Ediacarans may have been ancestors to the Burgess Cambrian animals or, more likely, because they are utterly unique, they may have been one of evolution's many "false starts." The earliest marine animals—whatever these were—may have fed on protoctists, including algae. The small size and relative mobility of these algae eaters probably guaranteed a nutritional niche without much competition. Only after animals began trying to eat one another—and evolving larger size and hard bodies

as defenses—did they become obvious in the fossil record. Mating, embryo-forming animals must have exploited resources for millions of years before they evolved hard, easily preserved parts. Cambrian fossils are only the tip of the iceberg of animal evolution. They were hardly "the first animals."

A thermodynamic truth is that as heat dissipates life organizes and its surroundings degrade. There is no life without waste, exudate, pollution. In the prodigality of its spreading, life inevitably threatens itself with potentially fatal messes that prompt further evolution. But sometimes waste can be fashioned into something useful.

Within protoctist or animal cells the concentration of calcium ions—Ca^{++} charged particles—is some ten thousand times less than that of sea water. Calcium phosphate becomes a fatal precipitate as "rocks" form inside cells when too much calcium enters. And by binding with phosphorus (as phosphate), errant atoms of calcium deprive the cell of an essential ingredient for making DNA, RNA, and membranes. By contrast, in controlled, small quantities calcium ions can be an intracellular resource. Poison depends on dose. In small doses calcium ions signal; indeed, they are part of the electrochemistry of thought. But the calcium excess must always be shunted outside of the cell. If the process of extruding calcium beyond cell frontiers lapses, then chemistry takes over. As people with kidney stones know only too well, calcium phosphate, where it doesn't belong, causes trouble. While animals remained soft they released calcium into seawater. Near the beginning of the Cambrian period in the Phanerozoic eon, however, some began to control their calcium extrusion.

Evolving, early animals converted potentially threatening blockages into living architecture. Our bones and skulls, like those of the fish-headed amphibians that preceded us, are still made of calcium phosphate salts (Ca_2PO_4). Some animals, such as corals, fashioned calcium, phosphates, and carbonates ($CaCO_3$) as their exteriors. Other organisms deposited calcium inside as teeth. Replacing organic cartilage, hard calcium phosphate infiltrated the proteins to provide the structural framework (shells and bones) attached to muscles. As armor emerged in some Cambrian creatures, teeth and poking appendages evolved in others to penetrate that armor.

Human industry has no monopoly on hazardous waste. Earlier life forms tell us by their examples that long-term survival involves not so much halting pollution as transforming pollutants. Termites build nests from feces and saliva. Pollution in the form of calcium excrement, patched and reworked by the busy muscles of animals, formed the basis for the first shells.

Evolutionary Exuberance

Becoming numerous, using their powers of movement to colonize new territories, animals evolved, thus creating and inhabiting new ecological niches. The Russian-American novelist and respected entomologist Vladimir Nabokov once suggested that patterns of certain butterflies look more like the whimsical touches of an artist god than traits that had merely blindly evolved. Yet with a fuller, less mechanical view of evolution such animal traits can be explained.

Evolution is no mechanical law but a complex of processes, sensitive and symbiogenetic, in part resulting from the choices and actions of evolving organic beings themselves. Natural selection is often said to "favor" this or that trait. But the nature that selects is largely alive. Nature is no black box but a kind of sentient symphony.

Of all the organisms conceived, spawned, hatched, and born, very few survive. Tastes in food and mates lead certain beings to generate more offspring than their fellows produce. In some cases females help designate the genetic makeup of populations by choosing the healthiest, most ostentatious, or strongest males. But conscious intervention in evolution by those so evolving can be of a more subtle variety. A butterfly's wing with an imitation raindrop showing a line displaced as if it had been refracted through water does not require a conscious author, and yet it could arise from consciousness. Beguiling artifices might arise from misperception by intelligent actors, for example, birds consistently mistaking an insect's wing pattern for a leaf. Nature is made partly in the image of mind.

Nabokov was right to say that the greatest enchantments in both art and nature involve deception. The element of surprise is the revelation that a given phenomenon of the environment was, until this moment, misinterpreted. Animals who experience surprise as a pleasure are likely to recognize camouflage and leave more offspring than are their less perspicacious brethren. Selection as nature, filled with live, sensitive beings is by no means blind. [PLATES 49 and 50]

Deception is very important in animal societies—so much so that some sociologists speculate that human technological intelligence is an evolutionary offshoot of "Machiavellian" social intelligence—the

49
An angler fish with
bioluminscent spots and
ribs. Phylum: Chordata.
Kingdom: Animalia.
The family (ceratioids) to
which this fish belongs
has bioluminescent
members more capable
than we of culturing pure
strains of glow-in-the-
dark marine vibrio
bacteria in their bodies.
These deep sea fish use
their bioluminescent
symbiotic organ to lure
unsuspecting potential
prey who mistake the
protruding appendage
for a small edible fish.

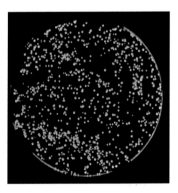

50
Photobacterium fischeri,
a bacterium. Phylum:
Omnibacteria. Kingdom:
Monera. Petri plate
showing colonies of lumi-
nescent bacteria. Many
types of fish cultivate
symbiotic bacteria in
special organs and put
their light to good use
in predator defense, food
illumination, or mate
signaling.

capacity to procure food, mates, child care, and so on by outwitting others in the tribe. The outwitting, outrunning, or outfighting need not be entirely conscious. Fearful apes and monkeys undergo a physiological reaction that stands their hairs on end. The effect to a potential combatant is an increase in size of the opponent, commanding if not fear then respect. The B-52, punk, and other "big hair" hair styles may similarly affect onlookers.

Since as naked primates we humans reveal more skin than hair, our own hair-raising response is rather measly: a prickly feeling, a hot flush across the nape of the neck, a tingling along the spine. Gooseflesh seems to be the evolutionary vestige of follicles that still go through the motions. Gooseflesh nevertheless exemplifies the evolutionary link between body and mind. A goosebump-mediated increase in a mammal's apparent size would be useless in a senseless world. But we live in a sensuous one, where details determine food and mate choices that in some cases spell the difference between life and death, between procreation and barrenness.

One of the sublime mysteries of life—supposedly calling evolution into question by its very presence—is the eye. Darwin wrote of the eye's "illimitable perfection." The eye, connected to the brain, perhaps seems perfect because it is the evolutionist's principle tool. But how might the eye, this subtle source of perspectival and reflexive mystery, have evolved?

The problem at first sight seems impossibly difficult. But not if we remember the microbes. Vision was anticipated in light-sensitive bacteria. Rhodopsin, "visual purple" of the mammalian retina, is a colored protein complex present in abundance in the pink, salt-loving archaebacterium *Halobacter*, where it is equally sensitive to light. The colored pigment portion of rhodopsin is retinal, a chemical similar to carotene of carrots and formed by the oxidation of vitamin A. Retinal, the absorber of light in the retina of the mammalian eye, has a 4,000-million-year history.

Using a plastid inherited from a cyanobacterial ancestor, the dinomastigote *Erythrodinium* functions as a kind of single cell eye. With its "imitation lens" and "imitation retina," this protist evolved a light-sensitive focusing device that involves most of its tiny body. Insects, flatworms, sea slugs, and frogs have eyes that are very different from one another, but all have carotene-derived light-sensitive membranes, lenses, and movable parts that direct light signals to locomotory organelles (such as undulipodia) or organs (such as muscles). Some evolutionists suspect eyes evolved in more than forty distinct lineages of animals. In all, light sensing is connected to movement in some way so that, once signaled, the creature can respond.

Sight, and its organ, the eye, may seem miraculous. Nonetheless, eyes exist along a continuum of complexity that both precedes mammals and is lately surpassing them, in forms such as infrared detectors, radio telescopes, and satellite imaging technology. Light sensitivity, in the rudimentary sense, even antedates life itself: colored compounds react in highly specific ways to visible solar radiation.

The human eye bears the marks of its microbial predecessors in other ways. The "rod" and "cone" cells are incapable of mitotic division; they bear kinetosomes and short undulipodia, inherited from their protist ancestors. The bony socket in which the eye resides derives from the autopoietic necessity of recycling calcium waste. Over time, life becomes more organized, integrating chemicals and even waste into beings so sensitive they eventually begin to perceive their own condition.

Messengers

By the late Devonian the phyla Arthropoda, Annelida, and Chordata had evolved representatives able to survive the rigors of life on land. Living beings evolved in water. The bacteria and protoctista cells since their inception were bathed in fresh and salty water. Desiccation was a dire threat any land pioneer had to overcome. Evolution of terrestrial species was no meager accomplishment. But the evolution of land animals was a triumph not just for individual organisms and species; it was a victory for the biosphere.

Movement and intelligence permitted land animals to act as vectors and messengers, to spread themselves to once-remote regions. By the early Tertiary birds had begun distributing phosphorus, a limited resource, to northern lakes and alpine peaks—simply by eating in one area and excreting in another. Carrying archaebacteria, ciliates, and other microbes in their rumens, cowlike animals digested grass, releasing methane, a greenhouse gas, into the atmosphere. Nitrogen-rich animal excrement accelerated algae growth, and fed fish and copepods in cold-water ecosystems. Especially during the Cenozoic, the most recent 65 million years, the quick reaction times, continent-crossing migrations, and complex social interactions of animals have accelerated activities within the biosphere.

But long before the Cenozoic, life was also a geological force. Photosynthesizing blue-green bacteria retained water in soil and sand, making Earth's surface green with chlorophyll. Carbon-precipitating life sequestered more and more carbon into coal and limestones, converting a tepid planet into one that indulged in glacial episodes. Land life created soil from planetary rubble. Ocean life transformed salts into reefs and evaporite flats.

"Proprioception" is the word for an animal's sensing various parts of its own body. Human beings today, likely the most populous of all mammalian species and certainly the most widespread, together behave as a kind of planetary proprioceptor, giving the biosphere sensations of itself. The greatest diversity of life exists in tropical jungles, such as the Amazon rain forest. Considering that distinct types of bacteria have merged to form the eukaryotic cell, and that colonies of eukaryotic cells evolved into animals, one wonders what may result from interaction of dense animal-rich communities.

Just as animal flesh was honed from the raw material of bacteria over eons, so complex interactions produced fledgling individuals at a scale beyond that of animals. Ants, termites, and bees form societies that carry out works in common. Reminiscent of human civilization, these arthropod workers methodically care for the young and divide labor among specialized castes of soldiers, workers, and reproductives. But whereas human civilization is only several thousand years old, fossil evidence shows that ants and bees have been organized into collectives for at least forty million, and termites for perhaps two hundred million years.

Together, animals confer their powers of movement and perception on the biosphere, making it an organized collective, the largest organic being of all. The animal actors of the global hive are at least 250 million years old. Snakes sense infrared radiation. Whales hear ultrasound. Bees detect the plane of polarization of visible light. Wasps see ultraviolet light patterns in flowers that look unpatterned to us. Dogs enjoy "ultrasmell." Sharks ferret out buried prey by detecting electrical potentials from the heartbeats of the hidden. Animals signal, and sense, and engage each other and their living environment in the visible, auditory, olfactory, and invisible radiative realms. Such sensitivity, so widely dispersed, sensitizes the entire biosphere.

Humans have extended one version of animal sensitivity into near Earth orbit. The image of Earth from space expands our awareness of the global environment. From the rudiments of animal sensibility and movement have come technological instrumentation, wheeled vehicles, and telecommunication. Together, the eyes of blackbirds, the sonar

51

Eschiniscus blumi, a "water bear." Phylum: Tardigrada. Kingdom: Animalia. These micro-scopic animals, named water bears by English nat-uralist Thomas Huxley are known as tardigrades. They survive drying out in temperature ranges from 150°C to -270°C. These microbeasts occur all over the world, but because the largest are no more than 1.2 mm in length, they remain obscure. The span from claw to claw in the photo is less than 0.5 mm.

of bats, the heat absorption of worms, the bacteria-derived luminescence of marine fish, and the aggregate awareness of untold numbers of walking, crawling, flying, burrowing, thinking beings produce more than the sum of the parts. Sensi-tivities interact. There are responses to responses. Animal awareness is not only a straightforward accumulation of eyes, ears, touch, and other senses but an incalculable synesthesia of mixed senses whose wholeness can be but gleaned by the human consciousness which forms only a part. [PLATE 51]

Both the French paleontologist-priest Pierre Teilhard de Chardin and the Russian atheist Vladimir Vernadsky agreed that Earth is developing a global mind. This layer of thought in the shape of a sphere they called the noosphere, from Greek noos, mind. The aggregate net of throbbing life, from flashing fireflies to human e-mail, is the developing planetary mind. Perhaps, like the brain of a human babe with many synaptic connections that diminish over time, the noosphere is still in its infancy. Polymorphous, paranoiac, confused, yet intensely imaginative, the thinking layer of Earth that is large-ly the unexpected product of animal consciousness, may now be in its most impressionable stage.

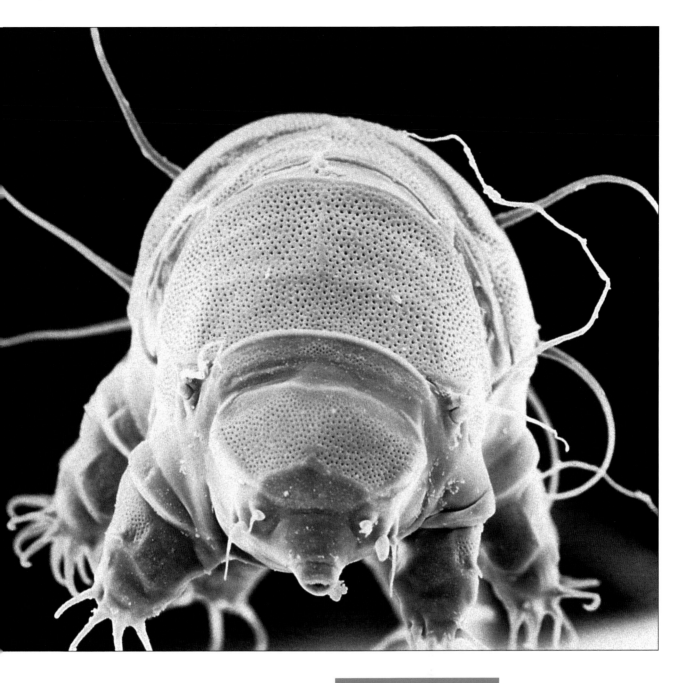

So, what is life?

Life is evolutionary exuberance; it is what happens when expanding populations of sensing, active organisms knock up against each other and work things out. Life is animals at play. It is a marvel of inventions for cooling and warming, collecting and dispersing, eating and evading, wooing and deceiving. Life is awareness and responsiveness; it is consciousness and even self-consciousness. Life, historical contingency and wily curiosity, is the flapping fin and soaring wing of animal ingenuity, the avant-garde of the connected biosphere epitomized by members of Kingdom Animalia. ∎

7

Flesh of the Earth

I hold that the fruit of the Tree of the Knowledge of Good and Evil was Soma, was the *kakulj,* was *Amanita muscaria,* was the Nameless Mushroom of the English-speaking people. The Tree was probably a conifer, in Mesopotamia. The serpent, being underground, was the faithful attendant on the fruit.

—R. Gordon Wasson

Truffles…are to be called certainly nothing other than flesh of the earth. Best at spring and more often during thunder, they are said to arouse dying love.

—Franciscus Marius Grapaldus

The Underworld

Academics still often partition life into zoology, the study of animals, and botany, the study of plants. But what of pink molds, single-celled yeasts, puffballs, morels, and psychedelic mushrooms?

Fungi have been lumped with plants because they aren't animals. Medieval scholars working in a three-kingdom system suggested they were zombielike, half-dead forms straddling the mineral and plant kingdoms. Until quite recently the scientific term for fungi has been Mycophyta—from Greek *mykes* (fungus, akin to mucus) and *phyton* (plant). Although none photosynthesize, like plants, some fungi are rooted. But they are best classified as unique in their own, solely fungal kingdom: the Mychota. "And fungi were fungi," wrote the Japanese poet Jun Takami, "they're like nobody else on Earth."[1]

In the English-speaking world the prototypical fungus is a dark, dank toadstool, dimly associated with witches, smelly feet, and refrigerators, and generally to be avoided. "Fungi," declared the eighteenth century French botanist S. Veillard, "are a cursed tribe, an invention of the devil, devised by him to disturb the rest of nature created by God."[2]

Fungi do require sex to form the morel or the mushroom, but they can reproduce without it. Because they don't photosynthesize, they can live in utter darkness. Their vampiristic existence often requires them to do so—sometimes on rather scarce resources of food and water. Reversing the animal technique of taking in food and then digesting it, fungi digest food outside their bodies. They then absorb the nutritious particles through their membranes.

Fungi differ from all other life. [PLATE 52, *page 142*] Unlike plants and animals they form no embryos. They grow from tiny propagules, packages called spores. Upon moistening, the spores form threads, thin tubes, the hyphae. Yeast cells (used in brewing beer and raising bread) bud off single cells. Lacking the whiplike structures of undulipodia, neither single-celled nor multicellular fungi ever swim. Some, called by the fancy name laboulbenomycetes, indulge in fungal sex to form spores that disperse on insect legs. Spores of others attach to mammal fur, are sneezed out or drift in wind. When hitchhiking spores come to rest and sense moisture, hyphae begin to grow up, down, and sideways. Like plants and animals, fungi are made of nucleated cells. Like plants (but not animals) they possess tough cell walls. Fungus cell walls are made of chitin, a nitrogen-rich carbohydrate; plant cell walls are made of cellulose. Many fungi have passages in their cell walls that allow mitochondria, nuclei, and other organelles to move between cells. Some lack cross walls altogether and are more a growing mass of tubes than multicellular individuals.

Fungi break down dead and sometimes live bodies. For more than 400 million years they have been settling and growing on a huge variety of foodstuffs other organisms eschew. A few grow in the sea or underwater, but they are basically land lubbers. Fungi were among the first organisms to make use of terrestrial environments, enabling the development of many other land dwellers. That fungi prefer land was proved when scientists returned to *Alvin*, a sphere-shaped submersible vehicle whose mission to the bottom of the ocean ended when its lifeline to the mother ship snapped. Two years later an intact, if soggy, sandwich was discovered on board the sunken scientific vessel. In contrast, even inside a lunch box any sandwich would be devoured if it were abandoned for two years on a picnic table at Earth's surface. Fungal spores are found virtually everywhere in the air but do not grow on the sea bottom, perhaps owing to an overabundance of saltwater.

Converting waste and corpses into resources, making nutrients available to land life, fungi are invaluable to global metabolism. Fungi thrive amid nature's informal coffins as undifferentiated masses; each fungal thread is bounded, but the expanding organism as a whole—a system of tubes—has no clear borders. Fungi, with their threads penetrating food today but excised by environmental contingency tomorrow, are truly fractal organisms.

The fungal feature commonly called toadstool is really only the minuscule tip of an amorphous underground web of living threads ("hyphae"). A mold growing on bread or fruit demonstrates the typical fungus's lack of regular boundaries. The fungal thallus, an old word for plant tissue undifferentiated into root, stem, or leaf, is also called a mycelium or mycelial network. As the verb "dial" has come to mean "depress buttons on a touch-tone phone," so botanical terms—fruiting body, spore—applied to fungi persist despite their inappropriateness.

Lacking discrete borders, fungi sprawl. Consider the tree-root colonizer *Armillaria bulbosa*. This organic being locates fresh food by continuous exploratory growth. A fungal clone, able to recognize itself from other expanding subsurface fungi, *Armillaria* has territorialized the forest underworld with such inexorable fortitude that, after an estimated 1,500 years of growth, one sample growing beneath virgin coniferous forest in Crystal Falls, Michigan, now encompasses some thirty-seven acres. This individual fungus has an estimated weight of more than eleven tons. A single individual, its genes have been sampled and compared throughout and determined to be the same. Such genetic stability is impressive. Fire, forest succession, and changes in food availability, have isolated parts of the sprawling *Armillaria*, which, nonetheless, maintains its genetic integrity. [PLATE 53, *page 143*] Are the broken-off pieces separate organisms? Or should they be regarded as the dispersed limbs of a single subterranean being? Reminiscent of a Stephen King novel, the great chunks of biomass

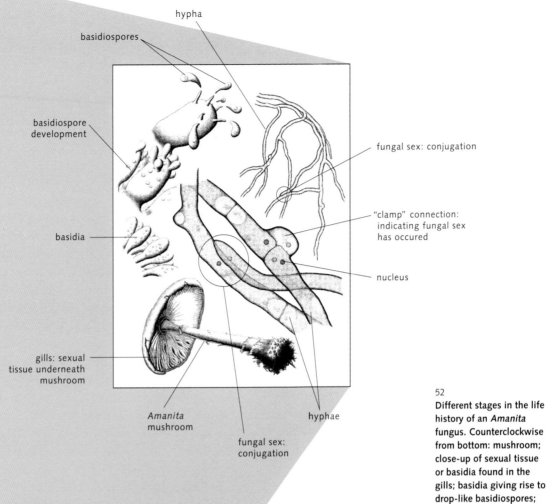

hypha

basidiospores

basidiospore
development

fungal sex: conjugation

"clamp" connection:
indicating fungal sex
has occured

basidia

nucleus

gills: sexual
tissue underneath
mushroom

Amanita
mushroom

fungal sex:
conjugation

hyphae

52
**Different stages in the life
history of an *Amanita*
fungus. Counterclockwise
from bottom: mushroom;
close-up of sexual tissue
or basidia found in the
gills; basidia giving rise to
drop-like basidiospores;
hyphae that grow from
spores; nuclei passing
through the hyphae in the
fungal sex act known as
conjugation.**

53
***Armillaria gallica* on malt
extract agar in a petri plate.
The fungal root-like hyphal
threads (rhizomorphs) can
be detected in a tangle in
the center.**

continue to thrive, unperturbed by their multiple amputations.

"Although fungi have acquired notoriety as agents of disease and as producers of psychedelic toadstools," writes scientist Clive Brasier, "their vegetative structure, the mycelium or fungal thallus, has a somewhat lower public profile."[3] The low profile is literal, as most of the mycelium exists in an extensive network, out of view beneath the soil. Great mycelia of foraging hyphae thrive beneath the forest trees. The living threads called hyphae tend to fuse. After "having sex," they eventually form mushrooms or mold tissues that in turn undergo meiosis, forming spores. These disperse over forest and field to grow again in search of mates.

Each mycelial network is a fungal clone, the far-reaching offspring of a single genetic line. Above ground, fungi produce air-borne spores, some of which you no doubt are inhaling now. When they land, the spores grow wherever they can. Sprouting networks of tubes, hyphae, into moist substratum, the fungi once again produce copious quantities of disseminating spores, spreading their strange flesh through the soil they help create.

Kissing Molds and Destroying Angels

No one knows how many species of fungi exist. Some say a hundred thousand. Others estimate one and a half million. Mycologist Bryce Kendrick of Waterloo University in Canada claims that today fungi are more diverse than plants but less so than animals.

As with the other four kingdoms—Monera (bacteria), Protoctista, Animalia, and Plantae—Mychota (fungi) have been arranged into phyla. There are five major groups or phyla of fungi. The zygomycotes (from Greek *zygon*, "twin" or "pair") or mating molds lack cross walls separating their cells. Mitochondria and nuclei travel easily through their open hyphal tubes. In zygomycote mating, special hyphae—the gametangia—grow toward one

another and fuse. From this fusing of gametangia comes the production of resistant spores. Once the hyphal ends lock, nuclei flow through the tubes and join, probably in pairs. When the mating is over, meiosis occurs, producing darkening spores in the head of a black spore holder. *Rhizopus stolonifer*, the most common of the black bread molds, is one example of a zygofungus.

Most molds (such as the pink bread mold *Neurospora*, *Claviceps*, and ergot fungi) and most yeasts (such as *Rhodotorula* and *Saccharomyces*) are ascomycotes. They form asci, sacs or capsules that develop when hyphae of compatible genders "kiss" and permanently fuse. The complex tissues and sexual spores produced from such liaisons is "mold." The hyphal threads, the nonsexual fungal body parts, are there but invisible to the naked human eye.

Ascofungi eat and degrade resistant plant and animal compounds, such as the cellulose and lignin of wood, the keratin of finger nails, and the collagen in mammal bones and connective tissue. By breaking down such compounds, these fungi release carbon dioxide, ammonia, nitrogen, and phosphorus to the rest of the biosphere. The evolution of wood placed great selection pressure on land-dwelling fungi to invent ways to degrade lignin, thus ensuring a coevolved biospheric cycling of matter. Some scientists posit that a lag in fungal evolution contributed to the worldwide accumulation of coal in the late Paleozoic era.

Humans everywhere are most familiar with members of the third fungal phylum. [PLATE 54] The basidiomycotes have reproductive structures called basidia, which resemble clubs (*basidion* is Greek for "club"). Familiar gilled mushrooms bear these spore-releasing basidia on their lower surfaces. Basidiomycotes include everything from the common supermarket mushroom (*Agaricus*) to the *Amanita virosa* (destroying angel) and its cousin, *Amanita caesarea* (a favorite of the Roman emperor Claudius). Among their numbers are to be found

54

Russula paludosa. Phylum: Basidiomycota. Kingdom: Fungi. This relatively common forest mushroom is connected to the roots of nearby trees with which it lives symbiotically.

55

Schizophyllum commune. Phylum: Basidiomycota. Kingdom: Fungi. The basidia of this mushroom are borne on the white double lines of the gills shown in the photograph.

giant puffballs (up to two feet in diameter), earth stars (looking like tiny breasts decorated with the leaves of a jester's cap), smuts, rusts, and jelly fungi [PLATE 55].

A fourth phylum, the deuteromycotes, consists of molds that form neither basidia nor asci. The deuteromycotes probably lost this ability when they lost sex. But they are nevertheless reproductive wizards, capable of the ceaseless production of airborne propagules. These organisms, also called conidial fungi or Fungi Imperfecti, reproduce by conidia, thin-walled cells that break off from the tips of ordinary hyphae. Others lack special reproductive structures; any part of the body, any hyphal thread or mycelial mass may break off to reproduce.

The fifth phylum of fungi consists of low-lying photosynthesizers — the lichens. Lichens are one of the most striking examples of symbiosis. They are also among the most successful fungi. Like the bacterial merger that led to algae, the lichen is a combination of fungus and alga (or sometimes, cyanobacterium). The result: an altogether new life form that takes advantage of the alga's ability to make its own food and the fungus's ability to store water and fend off the elements.

56

Cladonia cristatella, a lichen. A permanent merger between members of the fungus kingdom and photosynthetic members of the Protoctist kingdom, in this case the green alga *Trebouxia*. Thousands of different fungal species combine with a few photosynthetic partners to produce a great variety of lichens, such as this red-tipped British soldier lichen.

57

Isolated fungal partner of *Cladonia cristatella*, shown in the preceding Plate 56. The synergy of fungus and alga produces a structure that could not have been projected on the basis of simple addition.

58

Trebouxia, an alga. Phylum: Chlorophyta. Kingdom: Protoctista. The photosynthetic component of *Cladonia cristatella* seen in Plate 56.

Cross-Kingdom Alliances

All lichens—some 25,000 different kinds is an estimate—result from cross-kingdom couplings between fungus and either green alga or blue-green bacterium. Many lichens even harbor both types of photosynthesizers at once. Dwelling on bark and clinging to gravestones, sheer cliffs, and other sunlit places unavailable to less enterprising organisms, lichens have created a cozy niche for themselves. As they grow, they slowly turn solid rock inside out, into crumbling soil and living earth.

Divided, the fungal gray and photosynthetic green components of a lichen look nothing like one another. Nor does either member resemble their extraordinary composite. [PLATES 56, 57, and 58] The result of symbiosis, far from being predictable by simple addition, is a noncumulative surprise.

With so many different lichens, each representing a permanent tryst between fungus and photosynthetic life form, the phrase "long-term relationship"

takes on new meaning. When the photosynthetic partner of a lichen is removed, sugar excretion by this pigmented partner stops and does not begin again even if extracts of the sugar-inducing fungus are reintroduced. Somehow, algae and fungi sense each other's whole-body presence as they form an enterprising, complex partnership that depends on the history of the relationship. Like animal cells, the algal and fungal cells in a lichen communicate metabolically. Unlike most animals, however, the size and shape of any variety of lichen is not precisely fixed and the extent of their tissue complexity is limited to one or a few tissue layers. Lichens, however, surpass animals in their longevity; individual lichens may be 4,000 years old.

The ultimate interkingdom alliance, although not yet a lichen, may be the one that exists today in Antarctica. Seventy percent of Earth's fresh water exists in Antarctica, but the water there is tied up as

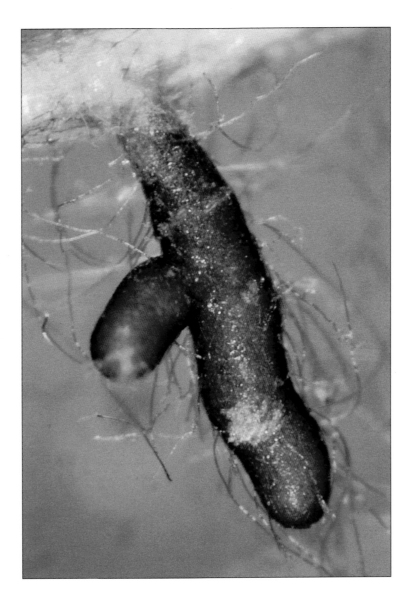

59

Mycorrhiza, a synergistic structure, a symbiotic protuberance produced by fungus and plant, in this case the fungus *Genea hispidula* on roots of the beech tree, *Fagus sylvatica*.

ice, and the relative humidity at these forsaken outposts rarely exceeds thirty percent. The few regions of Antarctica that are ice-free are thus deserts—and they are the driest places on Earth. Yet in the barren expanse of these cold deserts, so-named endolithic fungi grow with green algae inside the rocks. Feeding off the algae with which they loosely reside (and which obtain sunlight through the translucent rock crystals), endolithic fungi derive their water from rare but adequate melting frost.

Life can evolve suddenly, by jumps, when separate parties unite. Interkingdom alliances between fungi and algae produced lichens; a similar alliance may have been crucial to the development of the first forests. Root growths called mycorrhizae result from the dual growth of fungi and plants. [PLATE 59] Mycorrhizae provide the autotrophic plant partner with mineral nutrients, whereas the heterotrophic fungal partner is supplied with photosynthate food.

Galls on the stems of *Solidago,* goldenrod. Galls, "disease" structures, may represent symbiotic organs in an early phase of development. Such bulbous growths occur when plant tissue interacts with fungi, insects, and perhaps bacteria. It has been theorized that galls were the evolutionary predecessors to the first fruits.

Mycorrhizae—rounded, stubby, often colorful roots—are symbiotic, a dynamic structure produced by plant *and* fungus. More than 5,000 distinct mycorrhizal fungi have been discovered. Most of the associated plants seem to depend on these symbioses for their supply of soil phosphorus and nitrogen. Mycorrhizae look neither like plant root hairs nor like the fungus's mycelial network. They are synergistic, emergent structures, crucial to recycling. A single large tree may have a hundred different mycorrhizae, produced by distinct fungi, living in its roots.

Plants and fungi joined forces from the very start of life on land. Some of the oldest plant fossils in the world retain evidence of symbiotic fungi. Lacking leaves, branches, and twigs, the first land plants were little more than upright green stalks. Peter Atsatt, botanist at the University of California at Irvine, and Kris Pirozynski, former mycology researcher at the Museum of Nature in Ottawa, both contend that successful colonization of the land by the ancestors of modern land plants would have been impossible without root fungi. Today fungi are still synergistically intertwined in the roots of more than ninety-five percent of plant species. The algal ancestors of plants may not have been able to come ashore without nutrient-procuring fungi. The primeval arboreal carpet, the first forest floors, appear to have been created not just by plants but by plants and fungi acting together.

Kingdom Plantae is (and always has been) almost entirely terrestrial. The algal ancestors of this kingdom, of course, emerged from aqueous environments, but most of the descendants kept to the land. Early plants on land had to overcome almost insurmountable odds. Land then was mercilessly irradiated and depleted in the forms of nitrogen and phosphorus salts absolutely required for plant growth. Moreover, the land was and is an unreliable provider of that vital resource, water. Until the Silurian, when plants and fungi began to occupy

the land, cyano- and many other kinds of bacteria had the desolate continents to themselves.

Kris Pirozynski has hypothesized that fruits— whose colors, flavors, and aromas still cast an aesthetic spell over our primate brains—evolved by way of interferences from the fungi and animal kingdoms. His hypothesis attempts to explain the gap in the fossil record between the spread of flowering plants and the appearance of fleshy fruits a good forty million years later. Pirozynski envisions the first fruit appeared when fungal genes were transferred into plant chromosomal DNA. This is similar to what occurs in crown gall disease. Galls are symbiotic tissues formed by insect, bacterial, or fungal growth on plants. [PLATE 60] They are bloated, sometimes slightly monstrous-looking tumors found mainly on shrubs and trees—and some look remarkably like fruits.

Crown gall disease is caused in many plants by penetration of *Agrobacterium. Agrobacterium,* a kind of bacterium that lives in soil, bears plasmids (short pieces of DNA) that can enter into the cells within roots and stems of susceptible plants, bringing bacterial genes into the plant's nuclei. Biotechnology firms use *Agrobacterium* to introduce desirable genes into crop plants. Pirozynski speculates that fungal genes may have infiltrated plants in similar fashion. That fungal infections were important to the abrupt, relatively late appearance of fruits in many different flowering plants in the Cretaceous fossil record remains an intriguing hypothesis. That galls are examples of plant-fungal synergy, however, is an established fact.

Underbelly of the Biosphere

Fungi resemble animals in that, unable to produce food, they depend for nutrition on the bounty of others. From an ecological viewpoint, however, the two kingdoms differ markedly. Fungi are indispensable to the formation of soil, breaking down intractable rock. They help lay down the carpet of spreading life. They are the underbelly of the biosphere.

Without fungi, plants and eventually all animals would be starved for phosphorus (part of the autopoietic imperative as an essential component of RNA, DNA, and ATP). Fungi mediate the interstices of the food web. The Arab scholars who classified fungi midway between the plant and mineral kingdoms had a point. When fungi take over a body, its material nature is quickly revealed. Bodies become carbon-rich humus. Fungi decompose corpses and feed on living tissue, such as the skin of sweaty feet. For more than 400 million years their spores have been settling and sending mycelial networks through a global smorgasbord of foodstuffs. Recycling the dead, they are the garbage collectors of the biosphere.

Fungi break down bread, fruit, bark, insect exoskeletons, hair, horn, camera lens mounting compound, film, skin, building beams, cotton, feathers, the keratin of finger nails and scalp. Like portable cleanup crews, fungi are transported around the world by airborne spores. Almost nothing is exempt from their gastronomic ministrations. Indeed, their zeal for recycling is so great that many begin even before an organism has died. In maladies such as athlete's foot, jock itch, and ringworm, fungi get ahead of themselves in the job of redistributing elements of the biosphere.

Whether growing on human epidermal cells, breaking down the cellulosic fibers of cloth with cellulase enzymes to create mildew, or making green-gray spores as a penicillium mold begins its colonization of grapefruit, fungi digest substances left over by others. What seems like "decay" to us is, to fungi, the healthy growth of new offspring. Without fungi and bacteria breaking down complex macromolecules, the corpses of plants and animals would pile up, thereby taking phosphorus and nitrogen out of circulation.

Fungi, on land, perform most of the waste management function in the biosphere. Unlike ordinary people, who have survived as polluting nomads for generations, dumping and moving on, the biosphere cannot simply put its refuse out on the planetary curb. On Earth, garbage does not go out, but around. Humans are only now approaching the level of sanitary efficiency mastered 400 million years ago by the fungi: fungi do not simply remove garbage, they recycle it. Supplementing the bacteria, fungi recycle carbon, nitrogen, phosphorus, and the like; on a plant- and animal-dominated continental landscape they expanded planetary autopoiesis to dry land, changing Earth's surface forever.

Hitchhiking Fungi, Counterfeit Flowers, and Aphrodisiacs

As rapid recyclers fungi often send the rest of us mixed messages. Mediating at the interstices where garbage becomes food and corpses become fertilizer, fungi can cross up animal nervous systems.

Ever since amphibians and their descendants ambled onto land, we animals have had to contend with fungi. Indeed, for millions of years fungi and animals have coevolved. Our primate ancestors dwelled in forests and tasted many foods. Some were poisonous, others mind altering. When spores or hyphae, resisting digestion, can pass through animal intestines, it may be advantageous to them to be eaten. Those animals who find fungi delicious often offer them a nonstop free ride to the soil. For their part, animals that detoxified or vomited poisonous fungi survived to unwittingly disperse them.

As language evolved, social prohibitions against ingesting possibly dangerous fungi developed. So did sacred rites involving their use. The attempts of societies to try to rid themselves of drugs deemed threatening is reminiscent of the body's autonomic response to rid itself of certain fungal "foods." But due in part to the ability of certain species to produce mind-altering trips, fungi will never be completely eliminated from the body politic. Fungi are an entrenched part of sentient life in the biosphere.

In their ancient capacity as sanitary engineers fungi have evolved some rather startling relationships with members of other kingdoms. *Phallus* and *Mutinus* are penis-shaped stinkhorns of the order Phallales, whose stench, reminiscent of decaying meat, attracts flies. The flies that alight on them carry sticky fungal spores away on their legs. *Pilobolus crystallinus* is a kind of mating (or "kissing") mold whose favorite habitat is horse dung, rich in undigested cellulose, nitrogen, and other nutrients passed up by animals. Indeed, dung is so valuable to fungi like *Pilobolus* and other mating molds that they have evolved a clever stratagem to get there first: they arrange to be eaten, lying in wait amid blades of edible grass. Mature *Pilobolus* does not linger in the dung. Its sporeheads absorb water from the feces, internally building up intense pressures. Tensed for action, the sporeheads, unbranched structures two to four centimeters in height, aim toward the light. When internal

pressures exceed 700 kilograms/cm² the sporeheads pop out of the manure—and land several meters away, in grazeable grass. Here the long jumpers, an inspiration to the internationally acclaimed Pilobolus dance troop, can be eaten by helpful, munching horses.

Fungi have evolved other ruses. One yellow rust fungus, a basidiomycote, invented the insect-fooling stratagem of mimicking a flower. After infection the rust fungus induces rock cress (a plant in the mustard family that grows in Colorado's mountain meadows) to metamorphose and mimic the common yellow buttercup. Rock cress has unspectacular, droopy flowers but, once fungally infected, it elongates and makes a nectar-rich rosette—a rigid uplifted "flowerhead" that attracts would-be pollinators. Instead of pollen, the visiting insects pick up spores from these counterfeit flowers.

More than forty fungi species, such as *Panellus*, a delicate capped stalk, glow in the dark. Why they luminesce is not known, but a good guess is that in doing so they attract animals such as nematodes, tiny translucent worms. The worms eat the fungus, excreting and spreading the undigestible sticky fungal spores.

As ancient pioneers of land, fungi work with more recent settlers to propagate themselves and their offspring. The extraordinary willingness of fungi to perpetuate themselves by any means, and through any ends, possibly climaxes in the evolution of a particular fungus-and-ant agricultural system. Ever loyal to the fungi they cultivate for food, ants of the genus *Atta* evolved wheelbarrow-like depressions in their backs for carrying spores, which they then fertilize and feed with chomped-off bits of leaves, bark, and other plant material. [PLATE 61] These insects farm fungal spores as if they were seeds, carefully weeding out debris that might spoil their underground gardens. As with the fungal flower mimics, this cross-kingdom association brings to mind science fiction stories of beings enslaved to

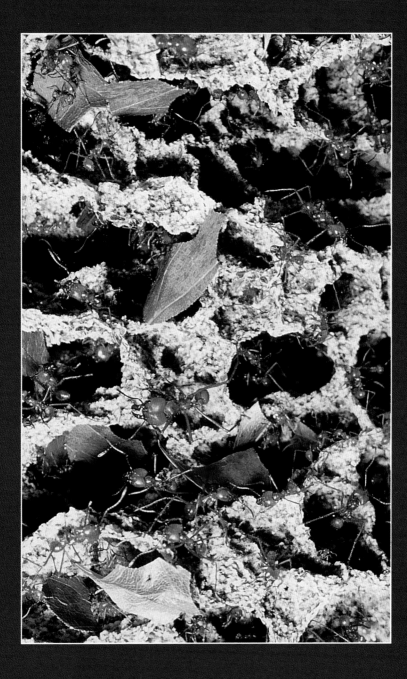

the dull task of propagating other beings. But in this instance, the ants, like harvesters enjoying the fruits of their agricultural labor, are fed for their efforts.

Human farmers have domesticated apples, bananas, and strawberries to the point that the sexually produced seeds are now infertile. To grow at all, the plants must be grafted or "vegetatively propagated" by human hands. Similarly, the ancestral sexuality of the basidiomycote fungi has been suppressed by long association with *Atta* ant farmers. In tending to their fungal gardens, ants have simplified fungal reproduction such that basidio-spore-bearing mushrooms, normally the outcome of fungal sex, no longer develop. Instead of mushrooms, bulbous structures called staphyla crop up as a result of all the plowing and planting. Like miniature fields of grain, rows of edible staphyla are harvested by the ant farmers as the major source of food for the colony.

The supermarket fungus, *Agaricus brunesscens*, is now grown in large limestone caverns in Pennsylvania. It has been thoroughly domesticated. ("It's like any other business," quips one executive familiar with commercial mushroom growing. "You keep 'em in the dark, feed 'em manure, and when they pop up you cut their heads off.") To gather Perigord black truffles, species name *Tuber melanosporum*, proverbial pigs (now more often dogs) are used to sniff the redolent hillsides of Provence in southern France and of Umbria in the Italian countryside. Although no farmers have yet been able to culture this delicacy, savvy mushroom lovers scatter truffle spores on the roots of oak seedlings before planting, just to raise the probability of finding mature truffles later.

Even the largest black truffles are fewer than three inches in diameter and weigh under two ounces. Truffles contain alpha-androsterol, a steroid that is also found in the breath of male pigs and thus may be the reason why sows were traditionally used to hunt the fungi. This steroid compound has been isolated from men's perspiration and the urine of women. It may be an ingredient not only in truffles but in the natural perfumes attracting the sexes. Rare as they are truffles would be even more so were it not for their ability to entice the sensibilities of certain mammals. For thousands of years truffles have been devoured by pigs, dogs, squirrels, people, and other mammals, who (inadvertently or not) are all quite happy to spread the spores throughout the woodlands in their droppings. Who knows? The mammalian habit of burying excrement—like burying the dead, of possible benefit to these odd beings—may be yet another fungal machination.

Hallucinogenic Mushrooms and Dionysian Delights

Truffles are not the only offerings made by Kingdom Mychota for the refined human palate. We savor gorgonzola and other blue cheeses, half-digested by molds. Nevertheless, we turn up our noses when other fungi invade milk products, declaring them "spoiled." Isobutyric acid lends its characteristic odor both to vomit and to some of the finest French cheeses.

Fungi are a diverse and wild bunch. Mushrooms convey something musty and strange and are often bypassed by foraging humans because of the difficulty of distinguishing edible from poisonous varieties. Indeed, it is at the interstices of edibility and toxicity that fungi produce their greatest manifestation: visions.

In the *Rig-Veda*, a Hindu sacred text, mention is made of drinking soma and of finding one-footed beings that live upside-down in the shade. These beings may have been mushrooms consumed for religious practices. The Eleusinian mysteries of ancient Greece also may have included the ingestion of fleshlike hallucinogenic mushrooms, perhaps *Amanita*, as part of a secret, drug-taking ritual.

In his written dialogues Plato records that Socrates, who left no writings of his own, gave the name pharmakon (both "remedy" and "poison" in Greek) to writing. The metaphor concerns writing's pharmaceutical nature: it expands memory, yet breeds dependency on writing implements. Even as writing expanded the storehouse of human knowledge, it weakened traditional story-telling skills. Socrates's argument against writing resembles the modern argument that television has a druglike effect on children and thus interferes with their learning (presumably more important) reading and writing skills. Or again, the comparison of writing to a drug is similar to the notion that electronic calculators should be banned because they prevent students from learning math for themselves. But hallucinogenic fungi are much older temptations than are writing, television, or hand-held calculators. They worked their strange magic on our fruit-foraging primate and pre-primate mammalian ancestors before anyone was in a condition to discuss it.

Because fungi feed on animal feces (and corpses), their interest often lies in making a rapid tour through the animal digestive tract; they entice animals to eat them but the spores and even the hyphae may resist digestion. Social mammals and birds serve fungi, transporting spores in their feces, sneezes, warm feathers, and fur.

Hallucinogenic fungi, such as *Psilocybes mexicana*, put a strange twist on the ancient relationship between mammals and fungi. The Catholic rite of the Eucharist, the flesh of God as mere wafer, may have originated in pre-Christian initiation ceremonies involving the ingestion of hallucinogenic fungi. Investment banker Gordon Wasson (1898-1986) even suggested that the "apple" Eve plucks from the Tree of Knowledge in the biblical Garden of Eden is actually a mistranslation of the word for an hallucinogenic fungus. An internationally renowned pioneer in the field of ethnomycology, the scholarly Wasson and his wife, Valentina, sought mushrooms in remote villages in Mexico during the mid 1950s. Their experiences watching and participating in the sacred mushroom rituals of the Mazatec Indians were recorded in a 1957 issue of *Life* magazine, and inadvertently, by informing the general public that such fungi existed, helped spark the psychedelic movement of the 1960s and beyond.

Wasson coined the term "mycophilic" to distinguish those cultures (notably in Russia, elsewhere in eastern Europe, Catalonia in northeastern Spain and southwestern France, and most of Korea and China) in which people are familiar and appreciative of fungi, from "mycophobic," or fungi-dreading cultures. Fungal ambivalence roughly parallels the divide along the former Iron Curtain. Save for Michigan and Wisconsin morel pickers and chanterelle fans of Washington and Oregon, the average American city dweller remains a borderline mycophobe no more likely to appreciate members of the Kingdom Mychota than to think of herself as a symbiotic permutation of bacteria.

Fungi provide an enchanting bevy of products, a veritable Dionysian garden of delights. They make the alcohol in champagne, wine, and beer; they raise bread and give it its fluffy texture; they ripen brie, camembert, troyes, thenay, and vendome cheeses; they flavor soy sauce and miso; they grow in the woods as delicious truffles, gourmet girolles, chanterelles, and morels. Oyster mushrooms, the saffron milk cap (shown on frescoes at Pompeii), the fairy ring mushroom (often dried and used as spice), the dangerous-looking bluish purple blewitt (sold in markets in northern England), shiitake mushrooms (*Lentinus edodes*), the enokitake or velvet stem mushroom, the nameko, the high-priced pine truffle or matsutake, and *Boletus edulis* (porcini to the Italians, Steinpilz to the Germans, seta to the Spanish, cepes to the French) all may be eaten with Rabelaisian relish.

Transmigrators of Matter

The Chinese elephant ear fungus, common in hot-and-sour soup, is thought to prevent cancer and retard heart disease. Fungi indisputably produce antibiotics. Englishman Alexander Flemming found quite by accident that a greenish mold, *Penicillium*, prevents bacterial growth. Penicillin, the active ingredient produced by this ascofungus, inhibits bacteria from making their cell walls, fighting strep throat, bacterial pneumonia, and other once-untreatable infections. Another mold, *Tolypocladium*, produces cyclosporine, one of the least toxic and most effective immunosuppressants known, used to prevent infections after organ transplants.

For every fungal evil—every refrigerator crisper invaded by a black stain, every retching resulting from an incorrectly identified mushroom, every peach gone the way of green fuzz—there is an equal or overcompensating fungal good. This book would not be possible without fungi: as nutrient scavengers fungi connect and feed the roots of trees, providing the wood and pulp precursor to this page. "Yes, happily language is a thing," writes the French literary critic Maurice Blanchot, "it is a written thing, a bit of bark, a sliver of rock, a fragment of clay in which the reality of the earth continues to exist."[4] Composed of dead and living things, a place where dirt comes clean and where wastes are reborn as hyphae and spores, the soil is largely a fungal phenomenon. Fungi are part of "the reality of the earth." They are the most resilient of land's eukaryotes, able to thrive on a great variety of compounds, break down complex organic molecules, and inhabit seemingly inhospitable domains, such as the inside of glaciers or the extraordinarily acid waters of Rio Tinto in southwestern Spain. [PLATES 62, 63, AND 64]

The lifestyle of fungi alerts us to the arbitrariness of our notions of individuality. Growing without discrete borders, fungi mate promiscuously with those of many complementary genders. Their tiny mated threads produce distinguishing organs of sexuality recognized as mushrooms, puffballs, shelf fungi, and the like. Many team up with plants, algae, or cyanobacteria to form composite organic

62
Rio Tinto ("red wine river") in southwest Spain at Huelva, famous since before the Roman Empire, has huge concentrations of minerals in exceedingly acid waters.

63
Most people believe that the mineral-laden water of Rio Tinto is barren of life.

64
Not so, extremely bizarre life forms, many like this obscure fungus *Hormonema pumorum*, survive in the iron-rich acid waters of this river.

beings: the mycorrhizae or the lichens. Planetary physiology—Gaia—is the result of interaction of innumerable beings, including the fungal networks. Gaia is symbiosis seen from space. Fungi cycle matter and turn waste into usable nutrients within a symbiotically integrated biosphere.

Any organism that appears or species that evolves at first has a chance. But to persist, life forms must survive not on their own but within a global environment. They become integrated, or they die away. As Takami attests, people tend to consider fungi freaks of nature, frightening and expendable. And yet, from a planetary perspective, fungi proved themselves long ago. For four hundred million years, from tropics to poles, they have been untiring recyclers. With spores riding the winds or hitchhiking on kin from other kingdoms, fungi have spread themselves about the globe, adventurous pioneers and desiccation resistors par excellence.

When animals die, fungi give them a natural grave. By way of fungi, corpses come to fortify grasses and trees. Sawed and crushed, their cellulose fibers become paper, books, immortal words or words to be recycled into more words. Reminiscent of the Eastern doctrine of the transmigration of souls, fungi are the transmigrators of matter.

Life creates. The global autopoietic system, Gaia, spins off creatures increasingly strange. For a while at least, even for millions of years, the global environment will tolerate bizarre sports, rapidly spreading pioneers, opportunistic monsters. But in the long run organic beings confront the limits of their own multiplication. They survive not alone but within a context of global life. Chomping locust swarms devour monocrops; guano birds defecate salts, moving phosphate and nitrate from sea to land. Rapid reproducers begin as infestations or infections but are tamed. Any rampant planetary population, any out-of-control "tumor," finds its economy. All growing populations integrate into the working biosphere or they become extinct.

In English the word fungus is virtually synonymous with an unwanted, surgically expendable outgrowth. Such a meaning might apply better to the hoarding, materialistic species we have become than to the organisms that nobly serve as biospheric undertakers, investing animal waste with life and turning corpses to soil.

Fungal spores may once have shared the air over land mostly with the propagules of far more ancient cyanobacteria and bacilli. In the long run, however, they were to intimately share the atmosphere with ameba and algal cysts, bacterial spores, fern spores, flowering plant pollen and seeds, and with flying insects, birds, and bats. They distributed themselves by ascospores, conidiospores, basidiospores, activated dry yeast, and as lichen propagules: sorelia and iscedia. Not only colonizers, they became powerfully selected degraders, recyclers, and agents of planetary redistribution.

Those of us who succumb to programmed death and whose remains are neither scavenged nor flamed go to a fungal underworld. The chemicals of our bodies are returned to the earth. Fungi keep life's compounds recycling. The global workings of Kingdom Mychota run as smoothly as a kidney or liver. We, with our nomadic past, are still adjusting to the idea that in a closed system the fruits of our labor and loins cannot accumulate endlessly. They must be distributed, returned to the system whence they came. This is a difficult lesson: the matter of our bodies, our possessions, our wealth is not ours. It belongs to Earth, to the biosphere and, whether we like it or not, that is where it is headed, again and again. Fungi help get it there.

So, what is life?

Life is a network of cross-kingdom alliances, of which Kingdom Mychota is a willing and crafty participant. Life is an orgy of attractions, from the trickery of counterfeit "flowers" to the strange allures of truffle and difficult-to-swallow hallucinogens. As fungi, life seeks out the underworld of soil and rot no less than the sunny vistas overwhelmed by photosynthetic beings. Life is self-renewing and fungi, as recyclers, help keep the entire planetary surface brimming with life. Transmigrating matter, molds and mycelia have found their calling. Creating and destroying, attracting and repelling, undertaking and overturning, they are part and parcel of *terra firma*.

The Transmutation of Sunlight

Tyger! Tyger! burning bright
In the forests of the night,
What immortal hand or eye
Could frame thy fearful symmetry?

—**William Blake**

If one has the patience, and the courage, to read
my book, one will see that it contains studies
conducted according to the rules of a reason that does
not relent . . . but one will also find in it this
affirmation: *that the sexual act is in time what the tiger
is in space.* The comparison follows from
considerations of energy economy that leave no room
for poetic fantasy, but it requires thinking on a level
with a play of forces that runs counter to ordinary
calculations, a play of forces based on the laws that
govern us. In short, the perspectives where such
truths appear are those in which more general
propositions reveal their meaning, propositions
according to which *it is not necessity but its contrary,
"luxury," that presents living matter and humankind
with their fundamental problems* . . . freedom of
mind . . . issues from the global resources of life, a
freedom for which, instantly everything is
resolved, *everything is rich.*

—**Georges Bataille**

The rays of the Sun determine the chief
characteristics of the mechanism of the biosphere.
They have completely transformed the face of the
Earth. To a great extent, the biosphere is a
manifestation of them. It is a planetary mechanism
which converts the radiations into new and varied
forms of energy, which completely changes the
history and destiny of our planet.

—**Vladimir Vernadsky**

Green Fire

The ultimate source for all life's energy, growth, and behavior is the sun. Burning like a cool green fire, photosynthetic beings transmute sunlight into themselves. [PLATE 65] Protoctists (coccolithophorids, diatoms, seaweeds) are the main transmutors in the sea; plants the main ones on land.

Plants represent a high point in bacterial coevolution. They have raised the biosphere to a higher dimension—up to a hundred meters from the soil surface. Yet they are newcomers to the photosynthetic guild. Plants have only dwelt on Earth for the last 450 million years. Evolving from algae, plants—almost exclusively land beings—went on to green the continents.

The blue whale, 26 meters long and weighing 180,000 kilograms, is the most massive animal ever to have lived, heavier by far than the largest dinosaurs. Nonetheless, next to behemoths of the plant world—such as the giant sequoia, who may weigh 2,000,000 kilograms—even whales are light. One clone of the quaking aspen, *Populus tremuloides*, is estimated to contain 47,000 trunks. Nominated by University of Colorado biologist Jeffry Milton as perhaps the biggest individual organism on the planet, this dispersed but connected tree covers 43 hectares in Utah. It is estimated to weigh 6,000,000 kilograms. [PLATE 66]

Books come from plants. So do boardwalks, oak desks, hashish, cotton shirts, chewing gum, coal, myrrh, clapboard houses, chocolate. Plants are the source of morphine, codeine, heroin, and other drugs similar to endorphins, pleasure-giving chemicals produced naturally in the mammalian body. Bark of *Salix*, the willow clan, gives us salicylic acid, aspirin; other plants make not only analgesics but astringents, antifungals, antispasmodics, pigments, caustics, cardiovascular agents, expectorants, diuretics, fumigants, hemostatics, insect repellents and toxins, perfumes, and anti-asthmatics.

Plants have been such a deeply embedded part of

65

A chloroplast, the intracellular structure which carries on photosynthesis. This close up, magnified 45,000x, was taken with an electron microscope. A hallmark of plant life, chloroplasts, as has recently been "proven" by genetic comparison, evolved from cyanobacteria that greened the world long before the origin of plant life proper. Algae and plants all seem to have evolved after larger cells merged with smaller ones. The large cells fed on but ultimately failed to digest the free-living cyanobacteria.

66

Populus tremuloides, quaking aspen. Phylum: Angiospermophyta. Kingdom: Plantae. A stand of quaking aspen in the San Juan Mountains, Colorado. A similar stand in Utah has been nominated as the largest "organism" on Earth. It has been so nominated by considering each genetically identical tree, the stem of a many-treed body. The aspens in the picture, turning color simultaneously, are not as extensive as the 43-hectare, 6,000,000-kilogram stand sharing a single root system in Utah.

the human environment that we now hardly notice them. Unless a bouquet of long-stem roses or box of chocolates arrives on the doorstep, plants are taken for granted. Even then, meaning or mood is evoked by the plant product as symbol rather than the plant itself.

Plant life presents us with an extraordinary richness of sights, smells, and tastes. The seasonal bursts of fragrant flowers have a beneficial psychological effect for dwellers outside the tropics; the mere sight of hills of undulating grass can produce serenity.

Of the nine recognized phyla of plants, only one has flowers. But that one phylum is so diverse that it is thought to account for more than half of all plant species. A full documentation of all species in the three hundred families of flowering plants would be such a monumental task that it has never been undertaken. "Such a listing," writes botanist Frits Went, "would have to describe about a quarter million known plants; to compile it, all the taxonomic botanists in the world would have to work together for years and years, and the finished product would have perhaps half a million pages, enough to cover a whole wall in a library."[1]

Plants were not, however, the first form of "green fire" on land. Fifty miles southwest of Las Vegas an 800-million-year-old fossil soil is preserved in rock. The carbon content indicates some kind of ancient photosynthetic life. From another spot in the American Southwest, eighty miles northeast of Phoenix, Arizona, even older samples of fossil soils were collected, corroborating a hypothesis by Susan Campbell and Stjepko Golubic of Boston University that photosynthesis on land began in cyanobacterial form 1,200 million years ago or earlier. Paleontologist Robert Horodyski of Tulane University in New

Orleans and geologist L. Paul Knauth at Arizona State University in Tempe contend that land was abundantly covered by photosynthetic microbes in the late Proterozoic eon.[2]

Not until the Silurian appearance of true plants, with their alternation of spore propagules (formed by meiosis) and gametes in sex organs (that fuse to make embryos), was there a full-fledged escape, probably at first seasonal, from the algal necessity of dwelling in water. [PLATE 67] Freeing itself of water, life on land evolved internal means of support and grew up on land. Land plants made a water-pressure system of structural support from the cellulose molecules found in bacteria and algae. Later, they evolved a stronger substance that, when combined with cellulose, would remain elastic but strong and supportive, even in dry conditions. This substance, lignin, is the chemically complex polyphenolic that gives woody plants their woodiness. With lignin, the biosphere began its vertical climb, extending life's realm to another, third dimension over the land. Biologist Jennifer Robinson has suggested that the great piles of coal left in the Earth's crust owe their existence to the lag after plants had invented lignin but before fungi had evolved means of decomposing it.

Just like other life forms, plants came from microbial predecessors. Their heritage is of photosynthesis, but they need not be photosynthetic. Some plants, even those with leaves and fruit, have abandoned the way of green fire, no longer photosynthesizing. Like those eyeless, subterranean mole rats that no longer see because they do not need to, some white plants have outgrown their dependence on direct sunlight. For example, *Epifagus* (beech drops) and *Monotropa* (Indian pipes) [PLATE 68, *page 162*] eke

67
Navicula cuspidata, a diatom alga. Phylum: Bacillariophyta. Kingdom: Protoctista. A diatom undergoing meiosis and gamete formation. Algae, like all protoctists, dwell in water. Plants eventually escaped water to venture onto land by evolving waterproofing such as waxy cuticles and structural support such as lignin.

68

Monotropa, Indian pipes.
Phylum: Angiospermo-
phyta. Kingdom: Plantae.
All plants develop from
embryos but not all plants
are photosynthetic. The
Indian pipes in this picture
are albino parasites that
have lost their chlorophyll
but have retained ghostly
remnants in the form of
blanched plastids.

Quercus oak leaf   acorn
(seed in fruit)

out nutrients by subvisible fungal threads to their
benefactors, neighboring green forest trees.

The distinguishing feature of plants is not
photosynthesis per se but that they all grow from
spores at one stage in their life cycle and from
embryos in another. The plant embryo, found deep
in the maternal tissue, is the diploid product of
sexual union. While it is, like an animal embryo,
formed from the fusion of a male propagule with an
egg held in a female organ (the "archegonium" in
plants), neither the egg nor the sperm of a plant is
produced by meiosis. Plant embryos form when
male pollen tubes or swimming sperm penetrate
tiny female plants or the female parts of hermaphro-
ditic plants. These tiny female plants are haploid
and grow from haploid spores buried in the mother's haploid tissue. The mature plant that grows from
the embryo does not make gametes (as mature animals do); rather, it makes, through meiosis in its
diploid body, spores. The spores grow into either
male or female haploid plantlets that make gametes
without meiosis. [PLATE 69]

Plants are sexual beings. Sexual coupling is the
act, and the embryo is the structure, that distinguishes them from algae and members of other
phyla (lichens, for example) that have sometimes

69
Different stages in the
sexual life history of an
oak tree, *Quercus.*
Counterclockwise from
top: oak leaves with two
mature fruits that contain
the seeds (acorns). The
flower shown at center
right is magnified and its
walls (the ovary) are cut
away to reveal the eight
nuclei of the double fertilization of angiosperms.
The pollen tube has pene-
trated the embryo sac and
let loose three small male
nuclei. One will fertilize

the egg nucleus (bottom
center) and one will fertilize two of the larger female
nuclei to form triploid
(3-chromosome set) tissue
which will nourish the
embryo—hence "double
fertilization." The doublet
anthers that produce the
pollen are shown on their
stalks at the lower right. At
the lower left a germinated
pollen grain that has
formed a pollen tube
is seen on its way into the
ovary of a bisected oak
flower.

been misrepresented as "plants." Plant sex, however, differs from that of animals: although fertilization of plant egg and sperm nuclei makes embryos, plant meiosis does not make eggs and sperm. Meiosis makes spores. Spores grow into plantlets, each with only a single set of chromosomes. These plantlets, called gametophytes, can grow—unlike an unfused animal egg or sperm. The gametophyte grows by mitotic division of its cells, which carry only a single set of chromosomes.

In cone-bearing and flowering plants the gametophyte (either male or female) is just a tiny structure that is not free-living. The gametophyte forms and spends its life entirely within a cone or flower of the plant that meiotically produced it. Plantlets make their sex organs and their gametes by mitotic cell division. Because they begin with only one set of chromosomes, these cells do not have to change their chromosome numbers to produce sperm or eggs. When the mating cells fuse, doubleness is reestablished and the cycle begins again.

Evolutionarily speaking, however, the failure of the gametophyte to "leave home," so to speak, is a recent condition. Plants of an older lineage, ferns for example, cycle through an alternation of generations in which the small bodies of those with a single set of chromosomes and the large ones with two sets are physically unconnected and strikingly different in form.

Overall, as sexual embryo-forms, plants and animals are more alike than either are like the three other kingdoms (bacteria, protoctists, and fungi). Animals are, however, diploids with a single-cell haploid phase, whereas plants have a multicellular haploid phase—and, unlike the cells of animals, all plant cells contain remnant cyanobacteria.

The Accursed Share

In the enigmatic epigraph to this chapter, Georges Bataille links the tiger with that initiating point of the mammalian life history, sexual intercourse.

He assures us, moreover, that his comparison is rational. It is. The entire unfolding of evolution is a response to an unexportable excess, a growing surplus of sun-derived energy. Both the sex act and the tiger are complexities of the biosphere.

While coitus is a behavior and the tiger a being, together they represent two fates of plants' prodigious reserves. The tiger is poised atop a pyramid of global nutrition whose base is the sun. Even at rest, the tiger represents life's nutritional edge, its carnivorous limit. The tiger, "burning bright" in Blake's memorable phrase, represents the funneling of solar radiation into a highly specific and potentially terrifying form. Coitus employs sun- and plant-derived wealth as animals expend energy to make more of themselves. Bataille further argues that classical economics is mistaken: the general economy is not human but solar. Sun-produced food, fiber, coal, and oil, carbon- and energy-rich reserves are the living foundation not only for bustling animal life but for industry, technology, and the wealth of nations.

The economy comes from photosynthetic life and the sun. Photosynthesizers use solar radiation to produce the cold hard cash of the biosphere. Heat is dissipated, degraded energy lost to space as primordial wealth accumulates. Colorful photosynthetic bacteria, protoctists, and plants the world over produce and "save." Eating them, consumers may "spend" through metabolic activities gathered photo synthetic energy or anabolically (and temporarily) store it in their herbivorous or predatory tissues. Primordial wealth may also end up in long-term storage (or be lost outright) when consumers die and are buried without decay.

Spending has always been a critical problem for life. Greed comes easily within a biosphere whose constituency triumphs as a function of ability to amass the wealth of photosynthesis. Bataille's tiger mercilessly hunts the leaf-eating deer. North Americans now fell trees to print paper money with colored fibers—or submit such bills in return for the

70

Forest floor in Coos County, New Hampshire. Fungi are here externally digesting leaf litter, animal corpses and other debris whose temporarily trapped, ultimately solar-derived energy will thereby be made available to other organisms in the biospere. In this way bacteria and fungi recycle and transport valuable compounds made of carbon, nitrogen, phosphorus, oxygen, and sulfur, spreading the wealth of the solar economy.

striped pelt of that endangered mammal. Photosynthesis creates excess, surplus, a reserve of matter and energy whose uses are as numberless as life is creative.

Bataille perceived that the character of a particular society is determined less by its needs than by its excesses. Wealth creates freedom in both biological and cultural realms. A nostalgia for old Europe, a respect for native American restraint, an admiration for the opulence of Egypt—these are sentiments based implicitly on the recognition that a culture is determined by how its members choose to spend or accumulate its excess. Rome makes its coliseum and basilicas, America its MacDonald's and Disneyland, Egypt its sphinx-guarded pyramids.

In the United States politicians grapple with tax collection, deficit and debt reduction, and public spending. The government prints money that banks lend without having or touching. Stocks, bonds, certificates of deposit, precious metals, and other instruments of finance are owned by investors. But what does it mean "to own"? Humanity does not own what it spends; ownership rests with the biosphere. Checks, credit cards, paper money, and stock certificates are all symbols of a wealth whose source lies beyond technological humanity's means of production. The monetary economy attempts to arrest the solar flux of Earth's economy. Money symbolizes the conversion of photosynthesis, life's energy, into something else—something that can be controlled, manipulated, and hoarded by humans. Perhaps it is no coincidence that in the United States money is green.

The fact remains that without plants the vast majority of animals would starve. Indeed, even with luxuriant plant growth humans and all other animals are destined to die. The grave is a great leveler, and a good reminder that we are owned by what we own. All of us from street sweeper to billionaire pay our dues. The elements of our bodies return to the biosphere whence they came. In the restricted economy of human arrogance and fantasy, individuals may amass great wealth and power. But in the solar economy of biological reality each and every one of us is traded away to make room for the next generation. On loan, the carbon, hydrogen, and nitrogen of our bodies must be returned to the biospheric bank.

A biosphere differs from an organism in that it is essentially closed to influx and egress of materials. Although supplies of carbon usable by life arrived with meteors and comets that penetrated Earth's atmosphere, especially before life took hold, this external source of material today is insignificant. Unlike an organism, eating and excreting, the biosphere has become self-contained. Its materials are limited, used over not up. The luxurious surplus of edible and usable compounds produced by photosynthesis leads to scavengers and predators, organisms killing and eating or cleaning up to survive and grow. The limited material reserves of the biosphere constrain the amount of solar rays that can be transformed into green life. [PLATE 70]

Overall, photosynthetic activity creates a surplus of energy-rich matter that can be hoarded, eaten for growth, or outright squandered. The great planetary

THE TRANSMUTATION OF SUNLIGHT

riches are there for the taking, replenished by the lively conversion of solar energy. It is an understandable but impossible wish to preserve the planet in its "original" state. The pristine nature to which some wish to return is not eternal but rather the green world that supported our ancestors so beautifully that they overpopulated it. Moreover, human spoilage of the lush environments that nurtured us is not evidence of any singular ability to imperil all life on Earth. No single species in the past has ever threatened all the others. Any tendency of one kind to overgrow and despoil was kept in check by all the rest. The essence of "natural selection" is that unstoppable tendencies of one population to grow to the point of environmental degradation will be halted by the growth of others. Human population expansion plays by the same rules: the degraded environment breeds morbidity, high mortality, and ultimately even extinction.

Our evolution has unearthed hoarded organic treasures, such as coal and oil to power cars and heat homes. Wealth in the biosphere ultimately comes from the sun. Organisms die, populations decline, and species become extinct. But the biosphere gets richer. Human burning of fossil fuels, for example, is exploited by plant life. Plants incorporate carbon dioxide released from this burning into their bodies. This is not to say that the current industrial mode of human living may not be dangerous or lead to increases in global temperature. Rather, the conversion to waste of a surplus by one life form has biospheric precedents: far from impoverishing the planet, the waste of one may, in fact, create more wealth for another.

In the strange solar economy individuals die, returning their bodily wares to biospheric circulation. Chemicals used in bodies are not lost. All organisms confront the combined difficulty and temptation of making use of that persistent photosynthetically-derived excess to which Bataille gave the name "the accursed share."

Ancient Roots

The first plants were probably like today's bryophytes. Mosses, liverworts, hornworts, and their kin lack the vertical stature of other phyla, owing to the absence of any system for fluid transport and hydrostatic support. Little more than masses of green cells favoring moist surfaces, bryophytes then and now lack leaves, roots, and seeds.

At the end of the Ordovician period the land surfaces were coarse, populated by low-lying cyanobacteria and soil algae but no plants. Where water was dependable, in rivers and lakes and along the borders of the sea, cyanobacterial mats became thick. In drier locales a sinuous, tough binding of blackish green soil particles and angular bits of rock covered the land. A modern analog of such terrestrial life prior to plants are the desert crusts of Utah, the Gobi desert, and the fields of Iraq. These crusts consist of cyano- and other bacteria, occasional algae, and fungi—all of which are quick to begin the green fire of photosynthesis (or revert to a quiescent state) when given moisture.

Some modern green algae ("chlorophytes"), especially the chaetophorales, have been proposed as similar to the ancestors of plants. Their chloroplasts contain chlorophylls *a* and *b*—the same pigments found in the chloroplasts of plants. Like the sperm tails of mosses and ferns, two quite different sorts of plants, motile chaetophoralean algal cells bear two undulipodia. These green cells have intercellular connections, plasmodesmata, that resemble the perforations through the cell walls of plants. Animal cells join by strengthening contacts utterly unlike the perforations of plasmodesmata of algae and plants. The details of their mitotic cell divisions and their walls made of typical plant cellulose suggest that certain chlorophytes, such as the modern filamentous green alga *Klebsorbmidium*, resemble the imagined ancestors to plants.

Today ferns, ranging from fewer than three centimeters to over twenty meters in height, still reproduce by aquatic methods of egg and swimming sperm, which they shed into nearby puddles. Even so-called higher plants, such as the *Ginkgo* (a showy tree with fan-shaped leaves and stinking cherry-like cones, indigenous to steep slopes in eastern China), bear their ancient heritage in the form of undulipodiated sperm. The many tails of a single sperm in cross-section are undulipodia with the same 9(2)+2 symmetry of microtubule arrangement that is found in motility structures from algal swim tails and *Paramecium* cilia to bull sperm and the fine hair cells of the human lung.

The oldest well-preserved plant fossils are from black cherts at a quarry in Rhynie, a hamlet in Scotland. Geologists believe that the Rhynie fossils owe their superb preservation to periodic flooding from a nearby, silica-rich spring. The fossil plants, such as *Rhynia*, bear swollen roots, suggesting that fungi were already symbiotic with plant roots four hundred million years ago.

A form suggesting the most ancient members of the latest kingdom is still living today. This is *Psilotum nudum*, an obscure plant dwelling in greenhouses and in Florida, Pacific Islands, and other sunny climes. *Psilotum* is a system of stems, a mere bundle of green growing sticks. It releases spores into the air that produce sperm, which can swim in most soil films or across puddles. After fertilization the embryo gives rise to new shoot growth. Like a bryophyte, *Psilotum* lacks roots and seeds. But unlike a bryophyte, it has a vascular system and stands upright. Lacking leaves, photosynthesizing along its stems, it may resemble the earliest forms of plant life.

Modern mosses and liverworts also suggest the shapes of primitive greenery. These bryophytes overcame the algal dependency on a fluid surround by bringing water to land with them. The mere existence of leaves in mosses and liverworts gave them a big edge over the ancient psilophytes which, like *Psilotum*, were limited in their ability to gather light.

But, unlike vascular plants, the bryophytes never evolved structural support. Bryophytes to this day never grow more than a few centimeters high. They are vulnerable to the tricks of vascular plants that can easily overtop them, robbing them of sunlight.

Although no one is sure and the fossil evidence is scanty, many botanists believe that the simpler, more aquatic bryophytes evolved earlier than did the structurally more complex and dryness-resistant plants. Bryophytes are soft-bodied; their fossil record is decidedly poor. Modern bryophytes are utterly dependent on surface waters; they have no roots to scavenge for water down into the soil. But they are by no means fragile creatures limited to swamps, pond edges, river rocks, and waterfalls. Some live in areas of seasonal moisture, growing mainly during the wet season. Others, notably the ingenious sphagnum moss, are the sponges of the land. They are world-class water scavengers, capable of holding up to a thousand times their own body weight in water, storing it for dry times. A mound of sphagnum moss, moreover, employs the dead in the task of water retention. Only the surface of the mound is alive, but the moss corpses in the interior and lower reaches retain water for their descendants.

Most of plant diversification occurred in the vascular plants with tough bodies and conductive tissue. These organisms grew up, literally. Horsetails, for example, were among the first organisms to tower into the air. Modern horsetails, called "scouring rushes," have silica in their photosynthetic stems. They were used by European settlers of North America to scrub pots and pans. But these silica-stiffened organisms grew much larger in the past than do their descendants today. Ancient horsetails in primeval forests, in the Devonian period 410 million years ago, stretched up to fourteen meters high.

Primeval Trees
Rhynia-type plants seem to have evolved into many extant and now-extinct forms. The ancestral

71

Glossopterus scutum, fossil seed fern. One of the extinct groups of cycadofilicales whose trees made up the first forest over 225 million years ago before the evolution of the first dinosaurs. These ancient forests, crushed beneath the surface by floating tectonic plates, became coal.

vascular form probably gave rise to progymnosperms—an extinct lineage that branched off in one direction to become tropical seed ferns, which themselves later gave rise to the flowering plants. Another branch became the conifers that brachiosaurs dined upon and that lived on to survive meteor impacts and ice ages. The primitive *Rhynia*type vascular plant also diverged to become ginkgos, spore-releasing ferns, horsetails, and *Psilotum*. The branching talents of the original stem-maker thus came to enshrub and enforest the world.

A huge group of vascular plants, as important as the dinosaurs to the animal kingdom, have gone extinct. Known as the cycadofilicales or seed ferns, none of these trees—which looked like overgrown pineapples—are alive today. They were not ferns at all. Unlike modern ferns, they made conspicuously large seeds. These seeds (not directly related to modern seeds) were a major evolutionary innovation. Seeds can wait through a drought or cold spell. They can survive a lack of light. Seeds were as crucial to the dispersion of plants as water-tight eggs were for the great diversification of reptiles.

Possibly the first plants to produce seeds, the cycadofilicales abounded 345 to 225 million years ago, before any dinosaurs. They were the makers of the earliest forests. Leaves of the genus *Glossopteris* (Greek for "tongue-leaf") are common fossils in rocks deposited in the southern supercontinent of Gondwanaland. [PLATE 71] Exposed to powerful tectonic forces, Gondwanaland cracked and the pieces drifted apart on continental plates 200 million years ago. Those pieces are now called South America, Africa, Australia, India, and Antarctica. *Glossopteris*, like more than 99 percent of the plant and animal species recorded in the fossil record, is extinct.

Neither *Glossopteris* nor any of its once-successful relatives survived to inhabit the southern continents Gondwanaland has become. But, once upon a time, greening the world for over a hundred million years, forests of seed ferns swayed in warm winds from the southern reaches of Gondwanaland to the tropics of Laurasia, the northern supercontinent. Now, after 125 million years, the forests of Gondwanaland exist as a semipetrified, energy-rich plant refuse: coal.

By the end of the Devonian period and the start of the aptly named Carboniferous (Mississipian and Pennsylvanian) 360 million years ago, Earth was forested. Whether from Rhode Island, Edinburgh, or western Pennsylvania, coal of this age is replete with remains of leathery leaves, thick roots, and scaly bark. In the basement of the Biological Laboratories at Harvard University are "coal balls" that were hauled away from their sites of origin in Illinois and Kansas. Many are taller and, because they are spherical, stouter than a man. Chopping through them, or peeling off their surfaces with acid-treated acetate, reveals the source of ancient plant tissue: leaves, bark, roots, and flowerless sex organs hardly the worse for 290 million years of burial.

Measured by genera and higher taxa lost, the mass extinctions of the Permo-Triassic 245 million years ago were far more devastating than the better-known end-of-Cretaceous event that extinguished all the dinosaurs. A major factor in the Permo-Triassic extinctions may have been expansion of glaciers or a long period of profound cold—perhaps itself generated by a comet or meteor impact that darkened the skies with debris sent into orbit. Seed ferns were tropical plants. The seedlings of seed ferns and the trees themselves were vulnerable to bitter cold. Before all the seed ferns became extinct, however, at least one of their ancestors gave rise to plants that could withstand freezing temperatures—the conifers.

Conifer fossils are older than those of flowering plants. Fossil seeds of conifers are detectable as raised portions on the underside of female cone scales. Spruce, cedar, pine, and many other cone-bearing trees and shrubs alive today remain green all year long. So did many of their ancestors, adept at surviving arid, wintry conditions. The pollen of

Pollen tubes containing sperm nuclei germinate and grow down to mate with the female egg nucleus deep inside the "embryo sac" of the flower in the double mating event that occurs in all angiosperms: one sperm nucleus mates with one egg, and two other sperm nuclei mate with the egg's "sister nucleus" to form a tissue that feeds the growing embryo. The pollen grains with cytoplasm and nucleus are stained orange and the pollen tubes are stained white. The pollen grains (male) and the embryo sac (female) are the gametophyte generations of angiosperms and therefore, before fertilization, all their nuclei have one set of chromosomes. Having one set of chromosomes is termed "haploid" in biological terminology. In the human body only sperm and egg cells are haploid; plants, however, may produce entire bodies—intermediate parts of their life cycle—that are haploid.

conifers is wind-borne. Fertilization of conifers leads to formation of seeds in the shelter of female cones. [PLATE 72] This change, from the ancestral method of releasing delicate water-borne sperm or short-lived spores such as those shed from the underside of fern fronds and destined to make tiny gametophyte plantlets, permitted evergreen conifers to dominate lands of seasonal ice and snow or aridity, as they do to this day.

Floral Persuasion

In contrast to all cone-bearing, naked-seed plants, the flowering plants have encased seeds—the result of the growth of the flower's ovaries into fruit. More than a quarter million species of flowering plants inhabit Earth. The seeds of these angiosperms bear embryos and the female organs transform to encase them.

Humans have a special relationship with angiosperms. Our primate ancestors lived among African flowering trees and fed, in part, on fruits—which had evolved lucious colors, arresting aromas, and other tempting qualities insofar as such qualities seduced us into becoming involved in their reproduction. Mammals dispersed the encased seeds and, by defecating, enriched the soil where angiosperms would sprout. Ancestors more closely resembling modern humans no longer lived in trees, but they kept their nimble hands and binocular vision while living in a new landscape that was another cross-kingdom creation. The grassland savannas were the workings of angiosperms that invented a way to grow from the base rather than their tips. Savannas were equally the creation of the large herbivores, whose grazing killed tip-growing forbs and young trees, thus "naturally selecting" the grasses. Finally, the savanna would have been impossible without its recyclers: protoctists and bacteria which digested cellulose in enlarged fore- and hind-guts coevolved in the large mammalian grazers.

Even today our species has a special relationship

Papaver somniferum, a poppy. Phylum: Angiospermophyta. Kingdom: Plantae. The name of the common poppy means "sleep-bringing" from the effects of its acrid and narcotic juice. Angiosperm plants, whose evolution dovetailed with that of mammals, still cast their floral spells on human eaters, drinkers and lovers. E. O. Wilson's biophilia theory suggests that we have genetically-embedded patterns of emotional response to other life forms. The color, smells and tastes emitted by flowering plants captivate with all the power of their 100-million-year aesthetic legacy.

WHAT IS LIFE?

with angiosperms. Angiosperm grains, fruits, stems, leaves, and roots are our primary foods—directly, or indirectly by way of domesticated livestock. (The only exceptions are intensive fishing communities.) We surround ourselves with furniture that is often made from the lignin of forest angiosperms. Angiosperms have taught us pleasure in nurturing them in flower gardens. Even the image of a branching tree to explain evolutionary phylogeny is easy to understand in part because of our ancient familiarity with the growth patterns of flowering trees.

Charles Darwin called the origin of the flower an "abominable mystery." [PLATE 73] Beautifully fossilized flowers and seeds indicate that flowering plants appeared in the middle latitudes of the northern hemisphere at least by 124 million years ago in the mid Cretaceous. Thus, within sixty million years of the last of the huge, flowerless, coal swamp trees, flowers had evolved and spread. Crab grass, philodendron, mother-in-law's tongue, spiderwort, Indian corn, pumpkins, tulips, coconut palms, and willow trees are all representative of flowering plants. Although the Amazon rain forest contains a concatenation of flowering plants, as little as 10,000 years ago that rain forest extended to only two percent of its present area. Like mammals, flowering plants—especially florid tropical jungles—are

recent evolutionary phenomena.

Some plants have evolved such a high degree of interdependency with humans that they no longer survive in the wild. A graphic example of such intimacy is *Zea mays*, Indian corn. [PLATE 74] Evolved from teosinte, an obscure American grass, corn of many varieties now tower over people in season around the world. Simply by choosing seeds from the sweetest and most bountiful cobs year after year, people have rendered *Zea* entirely dependent. Unless removed and individually husked and planted by human hands or agricultural machinery, corn cannot reproduce. Absent this assistance, the kernels, trapped inside the fibrous covering, never sprout.

The "green revolution"—the vast increase in population due to development of agriculture and, subsequently, cities—is, from a biospheric perspective, a major success story for flowering plants. Like the *Atta* ants that tend their fungal gardens in rows, human ingenuity and resources—farm animals, fossil-fuel-driven tractors, fertilizers, irrigation, and biotechnological apparatus—have been devoted to maintaining the livelihood of our favorite plants. Our primate brains, evolving in a world of flowering plants, are still devoted to preserving and extending that verdant, nurturing world. Our attractions to

74

Zea mays, maize. Phylum:
Angiospermophyta.
Kingdom: Plantae. Colorful
maize kernels on cob of
Indian corn. Commercially
produced corn requires
husking by human hands
or machines. This requirement is testimony to
thousands of years of co-
evolution of human beings
and *Zea mays.* Today such
corn can not reproduce
without human agricultural
intervention in its life
process.

angiosperms are deep and instinctive — so much
so that bottled essential oils are sold as perfumes,
foods and drinks are artificially fruit-flavored, and
clothes and toys are dyed shades of red, yellow, and
orange — "warm" colors used in the original adver-
tising campaign by which plants produced attractive
and tasty incentives that enticed animals to do for
them much of the work of fertilization.

Plants, too, have animals to disperse their offspring. We eat grapes but spit out the seeds, disseminating the plant. Bitter seeds and the hard pits of
fruits would merit the adjective "clever" were they
the fabrications of an organism with a hefty brain.
Like colorful grocery packaging, bright and flavorful
fruits with inedible or discardable cores manipulate
the animal into collecting and spreading the offspring of the plant.

In an example of the growing intimacy among
the many beings cohabiting the biosphere, these
immobile, muscleless, and brainless beings —
plants — have succeeded in appropriating the very
powers of restlessness and active perception that
separate them from beings to which they are commonly presumed to be inferior — animals. Like the
symbiotic anastomosis of branches on the tree of
life, the merging of plant reproduction with animal
sensitivity and taste is a demonstration of life's

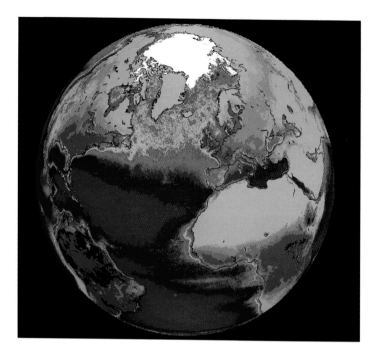

75

Satellite image of Earth from space showing the major zones of dominant vegetation: forest, desert, mountains, and other ecosystems. The continents here are seen as the raised portions of enormous moving plates that have changed their position over the history of the planet. Evolutionary biology and the new plate tectonic-based geology complement each other in broadening our view toward our living planet and its ancient history. A unified biosphere, Earth's surface is the chemically active aggregate of sunlight-transmuting, gas-exchanging, gene-trading, environmentally transformative life forms.

considerable powers of synergy and convergence. Living beings not only compete and struggle; they also associate and work together.

Solar Economy

We bipedal mammals like to think ourselves king of the Earthly hill, the most evolved form of life. But the argument might just as well be made in favor of flowering plants. They lack brains and speech — but, then, they don't need them. They borrow ours.

With our vaunted intelligence we have been Johnny Appleseeds, spreading fruit trees and grasses around Earth's surface. By tapping more directly than any previous animal into past and present photosynthetic powers, we raise the stakes of life on Earth. For, make no mistake about it, the solar economy has, with humans, entered a new phase.

Peter Vitousek, using satellite imagery, estimates that forty percent of the ice-free land surface of the globe is under agricultural cultivation; very little arable land remains untilled. [PLATE 75] Humanity annually uses the energy equivalent of 18 trillion kilograms of coal—about 3.6 metric tons for every man, woman, and child on the planet. This total energy is used, in part, to retrieve 327 billion kilo-

grams of iron, 90 billion kilograms of gypsum, and similarly staggering quantities of other materials. It is also used to generate and retrieve an estimated 540 billion kilograms of wheat, 90 billion kilograms of gypsum and 92 billion kilograms of seafood.

As fossil fuels and solar energy integrate into factory and machine production and into global husbandry and agriculture, more plants, animals, and microbes come to depend on the technological system now evolving. Nonrenewable resources are consumed, creating evolutionary innovations in the form of new biospheric waste: insecticides, polyvinylchloride, styrofoam, rayon, and latex paints. The gaseous byproducts of burning long-buried energy sources perturb or alter irreversibly the complex system of planetary physiology. Carbon dioxide accumulates in the atmosphere. Letting in visible light, but trapping reflected heat, this greenhouse gas may increase planetary temperatures—perhaps even melting polar ice, thus swamping coastal cities. Meanwhile, multiple extinctions follow from buzzsawing and bulldozing trees, killing some species directly but upsetting far more by destructive incursion into their living space. Nonetheless, the very energy our species uses to wreak habitat havoc comes ultimately from photosynthesis. For good or evil, novelty or status quo, nature is empowered by solar fire. The energy for violence also comes from plants.

Ever since *Homo sapiens* evolved, plants have fed, clothed, and sheltered us. From maternity ward vase to soft brown grave they accompany us on our biospheric journey. Borne on plants listing sunward, flowers symbolize peace, life, beauty, hope, femininity, and the sun.

Flowers, like tropical fish in an aquarium, elevate and calm. Fomenting biophilia, they are mental medicine, provoking our spirits. But flowering plants—all plants—are more than decorative. They are indispensable to an environment that can support humans. Their descendants will keep our descendants company. Spider plants, NASA reports, recycle trace pollutants in enclosed environments such as a space capsule. *Nymphaea*, a water lily, purifies drinking water. Long trips into space are inconceivable without plants to grow as food. Perhaps in seven generations your great, great, great, great, great grandaughter will look down at her toes—and see a wildflower poking up from a crack in the surface of Mars.

So, what is life?

Life is the transmutation of sunlight. It is the energy and matter of the sun become the green fire of photosynthesizing beings. It is the natural seductiveness of flowers. It is the warmth of the tiger stalking the jungle in the dead of night. Green fire converts wildly to the red and orange and yellow and purple sexual fire of flowering plants. Expanding, developing lignin, green beings raised up the biosphere and spread it horizontally. As fossils these beings trapped the original gold of the sun, stocking wealth only recently released in the human crucible of the solar economy. But the arrow in all these transformations must eventually become a loop that encloses the autopoietic exigencies of plants. We may be an intelligent life form but our very intelligence depends on that part of us we now recognize as photosynthetic. As life transmutes solar fire into all the energy and matter cycles of the biosphere, we pay homage to the ingenious ascension of the living plant. ■

9

Sentient Symphony

Owing to the imperfection of language, the offspring is termed a new animal; but is, in truth, a branch or elongation of the parent, since a part of the embryon animal is or was a part of the parent, and, therefore, in strict language, cannot be said to be entirely new at the time of its production, and, therefore, it may retain some of the habits of the parent system.

—**Erasmus Darwin,** *Zoonomia*

They say that habit is second nature.
Who knows but nature is only first habit?

—**Blaise Pascal,** *Pensées*

Thinking and being are one and the same.

—**Parmenides**

A Double Life

What is life? Two crucial traits are that life produces (autopoietically self-maintains) and reproduces itself. Then there is inherited change: DNA and chromosome mutation, symbiosis, and sexual fusion of growing life when combined with natural selection means evolutionary change. Nonetheless, autopoiesis, reproduction, and evolution only begin to encompass the fulness of life.

We have glimpsed ways of describing what life is: a material process that sifts and surfs over matter like a strange, slow wave; a planetary exuberance; a solar phenomenon—the astronomically local transmutation of Earth's air, water, and received sunlight into cells. Life can be seen as an intricate pattern of growth and death, dispatch and retrenchment, transformation and decay. Connected through Darwinian time to the first bacteria and through Vernadskian space to all citizens of the biosphere, life is a single, expanding network. Life is matter gone wild, capable of choosing its own direction in order to forestall indefinitely the inevitable moment of thermodynamic equilibrium—death. Life is also a question the universe poses to itself in the form of a human being.

Life is manifest on Earth as five kingdoms, each revealing from a different angle this mystery of mysteries. In a very real sense, life is bacteria and their progeny. Every available piece of real estate on this planet has become inhabited by subjects of the Kingdom Monera: by the enlightened producer, the tropical transformer, the polar explorer. Life is also the strange new fruit of individuals evolved by symbiosis. Different kinds of bacteria merged to make protoctists. When conspecific protoctists merged the result was meiotic sex. Programmed death evolved. Multicellular assemblages became animal, plant, and fungal individuals. Life is thus not all divergence and discord but also the coming together of disparate entities into new beings. Nor did life stop at complex cells and multicellular beings. It went on, forging societies and communities and the living biosphere itself.

Life is moving, thinking matter, the power of expanding populations. It is the playfulness, precision, and wit of the animal kingdom—which is a marvel of inventions for cooling and warming, moving and holding firm, stalking and evading, wooing and deceiving. It is awareness and responsiveness; it is consciousness and even self-consciousness. Life, historical contingency and wily curiosity, is the flapping fin and soaring wing of animal ingenuity, the avant-garde of the connected biosphere epitomized by members of Kingdom Animalia.

Life is the transmigrator of matter, in which task fungi serve as the closer of loops, making fungal food of plant and animal waste. Life thus seeks out the underworld of soil and rot as much as the sunny vistas enjoyed by photosynthetic beings. Life is a network of cross-kingdom alliances, of which Kingdom Mychota is an effective and seemingly crafty participant. It is an orgy of attractions, from the deceptiveness of counterfeit fungal "flowers" to the allure of truffles and hallucinogens.

Life is the transmutation of energy and matter. Solar fire becomes the green fire of photosynthesizing beings. Green fire becomes the red and orange and yellow and purple sexual fire of flowering plants, specialists in cross-kingdom persuasion. Fossilized green fire is hoarded in the human cubicle of the solar economy. Life is incessant heat-dissipating chemistry.

And life is memory—memory in action, as the chemical repitition of the past.

These halting descriptions approach but stop short of any final definition of life. We will not proffer any last word, final judgement because life will self-transcend; any definition slips away. In day-to-day adjustment and learning, in long-term action and evolution, in interaction and coevolution, organic beings go beyond themselves in the sense that they become more than what they were. Storing and redistributing the energies of the sun, life displays ever greater levels of activity and complexity. Who can guess what life might make of itself if and when it expands to remake a greater part of the universe into its home?

All organisms lead multiple lives. A bacterium attends to its own needs in the muds of a salt marsh, but it is also shaping the environment, altering the atmosphere. As community member it removes one neighbor's waste and generates another's food. A fungus goes about its business amid forest detritus, as it perforates the leaf of a nearby tree and helps close the loop in the biospheric flow of phosphorus. From one point of view we humans are ordinary mammals; from another, a new planetary force.

Like other animals, we eat, urinate, defecate, copulate; like them, we have descended from merged bacteria and meiotic fertilizing protoctists. Like other mammalian species, *Homo sapiens* should expect to endure for maybe another two million years or so—as the average species life span for mammals in the Cenozoic is fewer than three million years. All species disappear: they extinguish or diverge to form two or more descendant species. No animal species from Cambrian times is still alive today.

Perhaps *Homo sapiens* will diverge into two offspring species differing as much from us today as we differ from chimps, *Pan troglydites*. Such divergence may even be accelerated by technology. Human descendants, their nervous systems incorporated into durable robotic shells, may observe x-ray emissions of stars with telescope eyes as they cling to interplanetary spacecraft. Perhaps some ex-humans will free their genes of inherited disease or transcend—by genetic manipulation—normal intelligence. Others, dwelling on planets with higher or lower gravities, might undergo dramatic weight gains or losses, with altered bone mass and respiratory systems and rearranged internal organs.

Many scenarios are imaginable. But, whatever we become, our successors will retain traces of the past, which is our present. Consider: even if some new biological weapon were able to instantly vaporize all your animal cells, "you" would not disappear. As Clair Folsome mused:

> What would remain would be a ghostly image, the skin outlined by a shimmer of bacteria, fungi, round worms, pin worms, and various other microbial inhabitants. The gut would appear as a densely packed tube of anaerobic and aerobic bacteria, yeasts, and other microorganisms. Could one look in more detail, viruses of hundreds of kinds would be apparent throughout all tissues. We are far from unique. Any animal or plant would prove to be a similar seething zoo of microbes.[1]

We share more than 98 percent of our genes with chimpanzees, sweat fluids reminiscent of seawater, and crave sugar that provided ancestors with energy 3,000 million years before the first space station had evolved. We carry our past with us.

But now, aggregated into electronically wired cities, humans have begun retooling and transforming life on a planetary scale. Some futurists claim that we have become exempt from mere animal evolution. Are we not more than pretentious apes in fancy clothes? Do we not have music, language, culture, science, computer technology?

Rebuffed by Katharine Hepburn for a gin-drinking binge in the film *The African Queen*, Humphrey Bogart abdicates responsibility for his gruff behavior. "A man takes a drop too much once and a while," he explains, "it's only human nature."

"*Nature*," replies Hepburn primly, looking up from her *Bible*, "is what we are put in this world to rise a*bove*."

Self-transcending life never obliterates its past: humans are animals are microbes are chemicals. The view that we are "more" than animals does not contradict the materialistic perspective underpinning science. Life is less mechanistic than we have been taught to believe; yet, since it disobeys no chemical or physical law, it is not vitalistic. While we sense in ourselves a great degree of freedom, all other beings, including bacteria, also make choices with environmental consequences. Stored and transformed in life, the energy of sunlight powers cell growth, sex, and reproduction of highly similar life forms. All living beings may share our own feeling of free will.

Life on Earth is a complex, photosynthetically based, chemical system fractally arranged into individuals at different levels of organization. We cannot rise above nature, for nature itself transcends.

Nature does not end with us, but moves inexorably on, beyond societies of animals. Global markets and Earth-orbiting satellite communication, wireless telephones, magnetic resonance imagery, computer networks, cable television, and other technologies connect us together. Indeed, people already form a more-than-human being: an interdependent, technologically interfaced superhumanity. Our activities are leading us toward something as far beyond individual people as each of us is beyond our component cells.

Now, at the end of this century's hot and cold wars, we communicate freely across national boundaries at the speed of light, via telephone and computer. News flashes around the world. But these social changes at the start of the new millennium pale beside the sweeping biological changes. The Phanerozoic eon, which began more than 500 million years ago with widespread predation and defense against that predation in the form of animals with shells, is ending. Evolutionary movements that made eukaryotes out of bacteria and animals from protoctists are now repeating on a planetary scale. Humanity is transforming from a society into a new level of organic being. Our populations are beginning to behave as the brain or neural tissue of a global being. As we become more populous and sedentary, our human and technology-extended intelligence becomes part of planetary life as a whole.

The facts of life, the stories of evolution, have the power to unite all peoples. By integrating the data of thousands of scientists, and by cultivating the doubt and skepticism that is the epitome of scientific inquiry, the cultural invention called science could provide a more compelling, if ever corrigible, description of the world than do parochial myths and divisive, faith-demanding religious traditions. This hardly means that scientists are always correct. Yet the most meaningful story of existence for future humanity is more likely to come from the evolutionary worldview of science than from Hinduism, Buddhism, Judeo-Christianity, or Islam. The dual understanding of scientific inquiry and creation myth could become a single view: a science tale rich both in verifiable fact and personal meaning.

Choice

In a truly evolutionary psychology, spirit and mind are not celestial sprinklings but sovereign to living matter. Thought derives from no world but this one; it comes from the activity of cells.

When offered a variety of food stuffs, swimming bacteria, ciliates, mastigotes, and other mobile microbes make selections—they choose. Squirming forward on retractable pseudopods, *Amoeba proteus* finds *Tetrahymena* delectable but avoids *Copromonas*. *Paramecium* prefers to gobble small ciliates, but if starved for these and other protists it reluctantly feeds on aeromonad and other bacteria.

Although "merely" protoctists, foraminifera (one of the most diverse groups of fossil-forming organisms) make an astounding variety of magnificent shells. Foraminifera without their shells resemble amebas with very long, thin pseudopods. The shells are formed from sand, chalk, sponge spicules, even other foram shells. To appropriate their cell-shell homes, some forams agglute whatever particles are available from the environment together with an organic cement. Observations reveal, however, that, when presented with a hodgepodge of different particles, foraminifera make distinct choices based on shape and size—*Spiculosiphon*, for example, passes over much of the motley sediment, selecting only sponge spicules to make its test or shell.[2] Without brains or hands, these determined protists choose their building materials.

Smaller still, chemotactic bacteria just two microns long can sense chemical differences. They swim toward sugar and away from acid. A chemotactic bacterium can smell a difference in chemical concentration that is a mere one part in ten thousand more concentrated at one end of its body than at the other.

Biochemist Daniel Koshland explains the spiritual leanings of prokaryotes:

> "Choice," "discrimination," "memory," "learning," "instinct," "judgment," and "adaptation," are words we normally identify with higher neural processes. Yet, in a sense, a bacterium can be said to have each of these properties.... it would be unwise to conclude that

the analogies are only semantic since there seem
to be underlying relationships in molecular
mechanism and biological function. For example,
learning in higher species involves long-term
events and complex interactions, but certainly
induced enzyme formation must be considered
one of the more likely molecular devices for fixing
some neuronal connections and eliminating others.
The difference between instinct and learning then
becomes a matter of time scale, not of principle.[3]

Microbes sense and avoid heat, move toward or
away from light. Some bacteria even detect magnetic fields. They harbor magnets aligned in a row
in their tiny, rod-shaped bodies. [PLATE 76] That
bacteria are simply machines, with no sensation
or consciousness, seems no more likely than
Descartes' claim that dogs suffer no pain. That
bacteria sense and act, but with no feeling, is possible—but ultimately solipsistic. (Solipsism is the
idea that everything in the world, including other
people, is the projection of one's own imagination.)
Cells, alive, probably do feel. Indigestible mold
spores and certain bacteria are rejected by protists.
Others are greedily gobbled. At even the most
primordial level living seems to entail sensation,
choosing, mind.

Darwin formally distinguished "natural selection" (referring to interactions between nonhuman
life and its environment) from human-generated
"artificial selection" (the aesthetic or functional
choices of pigeon fanciers, dog breeders, and
agriculturalists). But "natural" selection is, in a
way, more "artificial," and far less mechanical, than
Darwin implied. The environment is not inert.
Self-awareness is not confined to the space between
human ears. Nonhuman beings choose, and all
beings influence the lives of others.

76
Electron micrograph of
magnetotactic bacteria
showing internal frag-
ments of magnetite. These
cells, able to orient themselves magnetically to the
north or south pole exem-
plify the sensitivity of living substance at all levels,
scales, and kingdoms.
Perception, choice, and
sensation apply not just to
human beings or animals
but, if they apply at all,
they apply to all life on
Earth.

Humans, we are told, are special. We have upright posture (allowing us to think of ourselves as literally "above" other species). We have opposable thumbs (man the tool user), linguistic abilities (man the storyteller), a superanimalistic soul (Descartes' distinction). We have, at least in the western hemisphere, a cultural tradition of seeing ourselves as being in a position of moral stewardship over the rest of life. Even in the absence of God, we imagine ourselves to possess a unique capacity to destroy the planet (via nuclear weaponry) or to swiftly change atmosphere and climate.

Even such an ardent foe of the idea of progress in evolution as Stephen Jay Gould (and he is not alone) proposes that whereas humans can evolve quickly through "cultural selection," all other forms of life on Earth are shackled to the ancient, plodding system of "natural selection." But the sheer number of traits listed to explain human uniqueness is enough to arouse suspicion. Among the dazzling array of reasons implying our superiority over the rest of life, one scientific argument stands out to us in curious contrast to the rest: humans are the only beings capable of wholesale self-deception.

This claim is based on early humans' presumably delusionary belief in the afterlife. Before written history our ancestors buried their dead with food, weapons, and herbal medicines of little use to corpses. How ironic that we, in seeking examples of our superiority over the rest of life, have finally congratulated ourselves on a trait that threatens to negate all the others! Although members of other species trick one another, humans are the expert self-deceivers: as the best symbol users, the most intelligent species, and the only talkers, we are the only beings accomplished enough to fully fool ourselves.

Little Purposes

Freud's understanding of unconsciousness as repression—painful memories are pushed away from the conscious mind—has diverted attention from another way in which actions become unconscious. Not avoidance but extended focus can make an action automatic, second nature. A speech is "learned by heart." A practiced typist no longer glances at the keyboard. Virtually any activity when memorized subsides from conscious attention. The heart pumps, the kidneys filter in autonomic quasi-perfection. Over breathing and swallowing, normally automatic, the willful organism can exert some voluntary control and modulation.

Now here is a strange thought: Perhaps we mammals remain unconscious of inborn physiology because under pressure of survival our ancestors consciously practiced their skills to unconscious perfection. Although modern science does not yet offer us a mechanism that transmits the learned habits of one generation to the physiology of the next, experience shows that conscious can become unconscious with repeated action. The gulf between us and other organic beings is a matter of degree, not of kind. Taken together, the vast sentience comes from the piling up of little purposes, wants, and goals of uncounted trillions of autopoietic predecessors who exercised choices that influenced their evolution. If we grant our ancestors even a tiny fraction of the free will, consciousness, and culture we humans experience, the increase in complexity on Earth over the last several thousand million years becomes easier to explain: life is the product not only of blind physical forces but also of selection in the sense that organisms choose. All autopoietic beings have two lives, as the country song goes, the life we are given and the life we make.

In the nineteenth century Samuel Butler (1835-1902), English author, painter, and musician, challenged Charles Darwin's account of evolution. Butler, who had many arguments with his father, left for New Zealand to become a sheep farmer after completing his education at Shrewsbury and St. John's College, Cambridge. Excited by Darwin's *Origin of Species* when he first read the book in New Zealand, Butler gradually grew disenchanted with it. A scholarly rebel who satirized society and explored the origins of religion, Butler accepted evolution but rejected Darwin's presentation. He began to suspect a dogmatism in the march of science as narrow as, but more insidious than, that of the church. Reading previous evolutionists, including Darwin's grandfather Erasmus, Butler accused the younger Darwin of failing to acknowledge his intellectual debts.

Darwin's schooling at Shrewsbury had been under the famous headmaster Dr. Butler—Samuel Butler's grandfather. The younger Butler claimed Darwin pretended to know little of his predecessors, and in the first editions of *Origin of Species* made it seem, upon returning from his famous Galapagos voyage, that the theory of natural selection had simply "occurred" to him. Butler pilloried Darwin's "Historical Sketch of the Progress of Opinion on the Origin of Species," which Darwin included in the second (1860), third (1861), and fourth (1866) editions of *Origin*. Darwin apologized for this entry described as "brief but imperfect." By the sixth and final (1872) edition the qualification was simply that it was "brief"—carrying the implication that the historical sketch had perhaps grown more nearly perfect in the interim. Butler disagreed.

What irked Butler most was Darwin's overly mechanical portrayal of the evolutionary process. Darwin, Butler quipped, had "taken the life out of biology." To be palatable to a religious populace in the Victorian era Darwin's evolution needed a credible scientific mechanism. Because the most respected achievement of the time was the physical science of Isaac Newton, Darwin portrayed evolution just as Newton had portrayed gravity: the result of abstract principles and mechanical interactions.

Although best known as author of *Erewhon* and the posthumously published *The Way of All Flesh* (an influential exploration of intergenerational struggle), Butler felt his greatest contribution was to evolution theory. Retreating from Darwin's neo-Newtonian presentation of organic beings as "things" acted on by "forces," Butler presented sentient life as making numberless tiny decisions—and thus responsible in part for its own evolution. Today Butler's view of the sum effect of little purposes escapes reprobation only where humans are concerned. We consider ourselves forward-thinking cultural beings, able to lay the flesh of the future upon the bones of imagination. We even believe we can govern evolution. The rest of life we dismiss as exempt from such Promethean foresight. Other organic beings are portrayed as the result of physico-chemical forces or unmediated genetics—too inert, by insinuation, to play a formative role in their own development. Butler begged to differ.

Butler's well-honed arguments and flashy polemics flouted the dry scientific prose of his Victorian day. One Butlerian theme stands out: living matter is mnemic, it remembers and embodies its own past. Life, according to Butler, is endowed with consciousness, memory, direction, goal-setting. In Butler's view all life, not just human life, is teleological; that is, it strives. Butler claimed that Darwinians missed the teleology, the goal-directedness of life acting for itself. In throwing out the bathwater of divine purpose, Darwin discarded the baby of living purposefulness.

No photosynthetic bacterium decided one day to become a willow tree. *Amoeba proteus* does not today set out to make itself into a mouse; it knows only that the swimming *Tetrahymena* it relentlessly pursues is tasty. Ameba-knowledge at this level of sensing and moving generates a million little such willful acts. These are sufficient for evolution to work its wonders.

Life's purposes are grandiose only in the aggregate, and in retrospect. Up close and confined in time they are ordinary. Nonetheless, no organic being is a billiard ball, acted upon only by external forces. All are sentient, possessing the internal teleology of the autopoietic imperative. Each is capable, to varying degrees, of acting on its own.

Butler's Blasphemy

In the second of four books published at his own expense, Samuel Butler discussed the evolutionary views of Erasmus Darwin, Jean-Baptiste Lamarck, Georges Louis Leclerc Buffon, and others. *Evolution, Old and New* (published in 1879) was thus a fitting title. In a circuitous series of newspaper letters and essays Butler published criticism of Darwin's work: he accused the great man of evading the credit owed Grandpa Erasmus and of mechanizing life. Butler even questioned Darwin's honesty. Butler, trying to put life back into biology, hoped Darwin would respond to his 1879 book or to his earlier (1877) *Life and Habit*.

Butler asked Darwin if his thought on evolution were an inspiration without precedent that had appeared to him from on high. Was it just the result of contemplation upon a great number of facts? Butler argued that the aura of poet-evolutionist grandpa Erasmus, coupled with far-flung reading of evolutionary ideas, must have also contributed to Darwin's intellectual development. Whether Darwin was a subtle master of self-promotion or

Butler a paranoiac of scholastic patience may never be resolved. But Butler was predisposed, by his own rebellious nature combined with Darwin's growing status as intellectual icon, to feel disappointment in the great man. A biographical book on Darwin's grandfather, Erasmus, had appeared. It was translated from German, and Darwin approved the translation as accurate. Butler read it and was alarmed to find that the translation into English of the early French evolutionist Lamarck used precisely the same words that Butler had used to translate Lamarck in Butler's own book, *Evolution, Old and New*. There was also a comment that those who wished to revive evolutionary thought prior to Charles Darwin showed "a weakness of thought and a mental anachronism which no one can envy." Butler believed he was being attacked obliquely in a fashion that precluded open debate; he believed that Darwin himself might have been responsible for the dismissal of the supposedly antiquated evolutionary ideas. He confronted Darwin first by letter, then in the newspaper. In a letter to Butler, Darwin explained that such alterations in translation as those made in the book about his grandfather were so common that "it never occurred to me to state that the article had been modified."

Darwin's family, and Thomas Huxley, urged Darwin not to respond to any Butler criticism or personal letters. He didn't. Yet biographical records show that he drafted two responses, never sent. In the second Darwin wrote he "could explain distinctly how the accident arose, but the explanation does not seem to me worth giving. This omission, as I have already said, I much regret."[4] Fascinatingly, Darwin vacillates even here, between the two modes of explanation Butler was to neatly summarize in one of his books as *Luck, or Cunning*: on the one hand, something was forgotten, there was a lapse of memory; on the other hand, there was an "accident," something was "accidentally omitted." Even here, close up, we see Darwin hedging, unable to characterize whether an event—and, indeed, one of his own doing—was the result of luck (chance, the "accident") or cunning (choice and design), the very theory Butler accused him of ignoring.

We agree with Butler that life is matter that chooses. Each living being, Samuel Butler argued, responds sentiently to a changing environment and tries during its life to alter itself. But living beings cannot effect changes with great efficiency. No light bulb appeared one day over the mammal that chose to become human. Rather, gradually, in tiny increments, living systems with nonnegotiable needs for food, water, and energy transformed themselves in wily and persistent ways.

What theologians called design, and considered otherworldly, was for Butler the result, in part, of Earth-bound thinking matter. The analogy of the writer comes to mind: she is blocked, with only a vague notion of what to write. Nonetheless, by following grammar, spelling, and syntax, by adding word to word, something purposeful emerges. The writerly result is not entirely her own because she complies with the rules of language. Similarly, no life flouts any law of physics, chemistry, or thermodynamics. As the decisions of writers exist in the lexical world, so the choices of living beings exist in the material world. Neither are absolute, and yet the deeper rules of matter, on the one hand, and language, on the other, impose structures that permit overall designs to arise, not perfectly complete but as accumulations of large numbers of minor, individual decisions.

Butler believed minute changes effected by organisms on their environment begin as conscious pursuits, but end as unconscious practice. For Butler, amebas too have their little wants, their little spheres of influence, their little "tool-boxes"

with which they materially change their environments, pursue their little goals, and build their little houses. This possibility is not ruled out by modern science. Nobel Prize-winning Danish physicist Niels Bohr (1885-1962), in discussing the striking utilization by organisms of their "past experience for reactions to future stimuli," contended that, despite the success of physics-based mechanistic biology, there was a need for description that includes "purposiveness."

> We must realize that the description and comprehension of the closed quantum phenomena exhibit no feature indicating that an organization of atoms is able to adapt itself to the surroundings in the way we witness in the maintenance and evolution of living organisms. Furthermore, it must be stressed that an account, exhaustive in the sense of quantum physics, of all the continually exchanged atoms in the organism not only is infeasible but would obviously require observational conditions incompatible with the display of life…it is evident that the attitudes termed mechanistic and finalistic [that is, purposeful] do not present contradictory views on biological problems, but rather stress the mutually exclusive character of observational conditions equally indispensable in our search for an ever richer description of life.[5]

For Butler living matter can "memorize" its behavior not only on the ontogenetic level of individual experience but also on the phylogenetic level of species history. The transition between ontogeny (the development of an individual life) and phylogeny (persistence and change of many individuals through time) is relative. The difference between the same individual newly born and at age eighty is greater, Butler argued, than that between a newborn infant of one generation and a newborn infant of the next.

Reptiles shed their skins; insects reshuffle their proteins in the pupa stage. The corpse is replaced, with some overlap, by its grandson. We moderns accept that a caterpillar "metamorphoses" into a butterfly, yet we employ the term "death" for what happens to grandfather's body. Nevertheless, with both the metamorphosing insect and the dying grandparent the fresh body of youth reappears. Each of us is entitled to think we die, suggests Butler, but the demarcation is highly arbitrary: a parent contributing to the flesh of a child is a prolongation, not an abrupt end, to biological continuity. The "individual" is not so complete, in time, as we have been taught.

Butler believed unconsciousness applies not only to adult human beings but throughout many different levels of living organization. The most important tasks, the ones most often repeated, have become the most unconscious, the most "physiological." The pumping of blood is too ancient and important to be unlearned or modified easily by mere lassitude or an act of will. Divorce from playful consciousness insures that routine but critical activities are competently performed. Treated as work, effected regularly and automatically, important physiology is not exposed to experimentation that might destroy it. The once-conscious process of steering an automobile recedes to the unconscious while the mind attends to other matters.

No concert pianist strives to hit the keys in order. The long periods of former conscious striving ingrain the knowledge into the musician's fingers. Dancers call their talent "muscle memory." Choice and practice become smooth habit.

Cells, with eons of practice, do not consciously decide to respire oxygen or reproduce by mitosis. They, or the bacterial remnants that compose them, may once have so strived. The more recent the addition of a habit to life's physiological repertoire,

the more likely it is to be conscious or at least subject to conscious interference. The metabolic pathways obtaining energy by conversion of oxygen reacting with food hydrogen into water and carbon dioxide probably have no congress with any of the degrees of sentience possessed by animals, plants, and fungi. Performed by mitochondria, which were once free-living bacteria, this chemical feat has not ceased for two thousand million years, ever since oxygen respirers responded to an environment transformed by oxygenic cyanobacteria. By contrast, peristalsis in digestion—also an unconscious phenomenon but one of which mammals tend to be aware—evolved in animal ancestors long after the microbial stage. Swallowing, chewing, and speaking are behaviors that have been learned far more recently, and in that order.

Butler's theory of unconscious memory holds that all beings are capable of forming habits, some of which—upon much repetition—become physiologically ingrained over the course of evolution. We do not remember, writes Butler, when first we grew an eye. Someday, he muses, so many of us will have learned to read and write, so often, that we will be born that way. Perhaps a future William Harvey will be required to uncover the details of how the learning of reading became a physiology, just as Harvey illuminated the circulation of the blood. Here we disagree with Butler. It seems doubtful that children will ever be born reading and writing. However, if one substitutes television for writing one can see that Butler is already on target: children are born, more or less, watching television; and television, put together from technological products made by companies all over the planet, is already so complicated that few humans understand it in detail.

Habits and Memory

Physicist Howard Pattee laments the simplistic misapplication of deterministic physics to biology. Pattee takes issue with biologists who are too quick to use classical physics to justify an understanding of life as a mechanical phenomenon. A general property distinguishing life from nonliving matter is its historical coherence, expressed as the potential to evolve. "Even though we do not understand the mechanism," Pattee continues, "the only conclusion I have been able to justify is that living matter has distinguished itself from nonliving matter by its ability to achieve greater reliability in its molecular hereditary storage and transmission processes than is obtainable in any thermodynamical or classical system."[5] The hints and hunches must be replaced by firm detail, yet the extraordinary storage and transmission processes of life for molecular heredity, mentioned by Pattee, may be robust enough to encompass the phenomenon postulated by Butler: phylogenetic "memorization," the conversion of the conscious strivings of one generation into the activities and, eventually, the physiologies of the next.

Although we fail as yet to see how an organism's or even a species' voluntary habits can become the physiology of a future generation via a material basis of heredity, we are fascinated by Butler's suggestion. We know, for example, that many organic beings acquire new heritable traits by symbiogenesis and that a vast array of others, not only people, are capable of learning. Ecosystems grow increasingly complex and sensitive; processes practiced in them repeatedly by one generation may become easier for the next. More open-minded investigation is needed. Objections may be leveled against Butler's ideas, yet he cannot be accused of the atavistic thinking which clings to humanity's separate status. Covertly considering ourselves divine, under the scientific rubric of "cultural evolution," or by dint of that other desperate euphemism, our "big brains," we are probably now more ecologically impoverished than we would be if, a century ago, we had adopted Butler's notion of all life as a conscious continuum.

Butler did not object to evolution but to the loss of the richness of the earlier, more lively views, in which living beings themselves were involved in natural selection:

> According to Messrs. Darwin and Wallace, we may have evolution, but are on no account to have it as mainly due to intelligent effort, guided by…sensations, perceptions, ideas. We are to set it down to the shuffling of cards.…According to the older men, cards did indeed count for much, but play counted for more. They denied the teleology of the time—that is to say, the teleology that saw all adaptation to surroundings as part of a plan devised long ages since by a quasi-anthropomorphic being who schemed.…This conception they found repugnant alike to intelligence and conscience, but, though they do not seem to have perceived it, they left the door open for a design more true and more demonstrable than that which they excluded.… They made the development of man from the amoeba part and parcel of the story that may be read, though on an infinitely smaller scale, in the development of our most powerful marine engines from the common kettle, or of our finest microscopes from the dew drop. The development of the steam-engine and the microscope is due to intelligence and design, which did indeed utilize chance suggestions, but which improved on these, and directed each step of their accumulation, though never foreseeing more than a step or two ahead, and often not so much as this.[6]

Existence's Celebration

For nineteenth century Englishmen of science it was natural and expedient to invoke Newtonian mechanics and conceive of life as Newton's matter: blind bits predictably responding to forces and natural laws. Like some piece of well-made clockwork, the world was donated or its mechanism manufactured by a transcendent god—a god that then stood outside its creation.

This was the new view of evolution: God, if it existed, was Newton's God. Not an active interloper in human affairs, it was the god of the mathematicians, the geometer god, who made the laws and then sat by and watched those laws play out. But an older view left room for a kind of god, too—a more active god. This was the view that Samuel Butler attempted to resuscitate—that life itself was godlike. There was no grand design, but millions of little purposes, each associated with a cell or organism in its habitat.

To the neo-Newtonians, the Darwinians, free will had been all but banished from the universe because the universe was portrayed as a mechanism and mechanisms do not have consciousness. For Descartes, God continued to have consciousness and people did insofar as they were in touch with God. But when Darwin showed through painstaking work that people too could be explained by the mechanism of natural selection, consciousness suddenly became redundant in the human world as well. Butler brought consciousness back in by claiming that, together, so much free will, so much behavior becoming habit, so much engagement of matter in the processes of life, so many decisions of where, how, and with what or whom to live, had shaped life, over eons producing visible organisms, including the colonies of cells called human. Power and sentience propagate as organisms. Butler's god is imperfect, dispersed.

We find Butler's view—which rejects any single, universal architect—appealing. Life is too shoddy a production, both physically and morally, to have been designed by a flawless Master. And yet life is more impressive and less predictable than any "thing" whose nature can be accounted for solely by "forces" acting deterministically. The godlike qualities of life on Earth include neither omniscience nor omnipotence, although an argument could be made for earthly omnipresence.

Life, in the form of myriad cells, from luminescent bacterium to lily-hopping frog, is virtually everywhere on the third planet. All life is connected through Darwinian time and Vernadskian space. Evolution places us all in the stark but fascinating context of the cosmos. Although something odd may lurk behind and before this cosmos, its existence is impossible to prove. The cosmos, more dazzling than any sect's god, is enough. Life is existence's celebration.

Butler's forgotten theory intrigues us. The mind and the body are not separate but part of the unified process of life. Life, sensitive from the onset, is capable of thinking. The "thoughts" both vague and clear, are physical, in our bodies' cells and those of other animals.

In comprehending these sentences, certain ink squiggles trigger associations, the electrochemical connections of the brain cells. Glucose is chemically altered by reaction of its components with oxygen, and its breakdown products, water and carbon dioxide, enter tiny blood vessels. Sodium and calcium ions, pumped out, traffic across a neuron's membranes. As you remember, nerve cells bolster their connections, new cell adhesion proteins form, and heat dissipates. Thought, like life, is matter and energy in flux; the body is its "other side." Thinking and being are the same thing.

If one accepts this fundamental continuity between body and mind, thought loses any essential difference from other physiology and behavior. Thinking, like excreting and ingesting, results from lively interactions of a being's chemistry. Organism thinking is an emergent property of cell hunger, movement, growth, association, programmed death, and satisfaction. Restrained but healthy former microbes find alliances to make and behaviors to practice. If what is called thinking results from such cell interactions, then perhaps communicating organisms, each of themselves thinking, can lead to a process greater than individual thought. This may have been what Vladimir Vernadsky meant by the noosphere.

Gerald Edelman and William Calvin, both neuroscientists, have each proffered a kind of "neural darwinism." Our brains, they say, become minds as they develop by rules of natural selection.[7] That idea may provide a physiological basis for Butler's insights. In the developing brain of a mammalian fetus, some 10^{12} neurons each become connected with one another in 10^4 ways. These cell-to-cell adhesions at the surface membranes of nerve cells are called synaptic densities. As brains mature over ninety percent of the cells die! By programmed death and predictable protein synthesis connections selectively atrophy or hypertrophy. Neural selection against possibilities, always dynamic, leads to choice and learning, as remaining neuron interactions strengthen. Cell adhesion molecules synthesize and some new synaptic densities form and strengthen as nerve cells selectively adhere and as practice turns to habit. Selection is against most nerve cells and their connections but it is nevertheless for a precious few of them. Of course, more work is needed before the physical basis of thought and imagination can be understood, but selective death in a vast field of proliferating biochemical possibilities may apply to minds as it does to evolution.

The peculiarly curved early embryos of birds, alligators, pigs, and humans are remarkably similar. Developing from a fertile egg, all display a stage with gill slits—whether the hatched or born animal breathes oxygen from air or water. The slits that close behind the ears in the human fetus attest to our common ancestry with fish, whose gill slits function in the adult. Human embryos have tails. Living matter "remembers" and repeats its origins to arrive in the present. In a Butlerian world, the materials of living beings are molded by life, over and over, for millions of generations. Creating a sense of *déjà vu,* the embryo represents a once-unconscious process, now again—at a different level—brought to consciousness.

Superhumanity

A transhuman being, superhumanity, is appearing, becoming part of the sentient symphony. It is composed not only of people but of material transport systems, energy transport systems, information transport systems, global markets, scientific instruments. Superhumanity ingests not only food but also coal, oil, iron, silicon.

The global network that builds and maintains cities, roadways, and fiber-optic cables grows by leaps and bounds. In Nigeria, for example, the population is expected to reach 216 million by the year 2010, double that of 1988. Unchecked, such growth would bring the number of Nigerians to more than 10,000 million by the year 2110—twice the present global human population. Our stupendous population taps a significant proportion of the solar energy reaching Earth's surface. The raw energy of photosynthesis, past and present, and transformed into edible plants, animal fodder, geological reserves, and human muscle and brain, supports the massive construction of the transcontinental urban ecosystem and even—"biting the hand that feeds it"—the razing of forests capturing and converting solar energies. [PLATE 77] As the system expands using genetic and atomic technology, its operations become more elegant and cohesive. The potential for disaster also increases.

Superhumanity is neither a simple collection of humans nor something other than aggregated humans and their devices. Plumbing, tunnels, water pipes, electric wires, vents, gas, air conditioning ducts, elevator shafts, telephone wires, fiber-optic cables, and other links enclose humans in a rapidly growing net. The way superhumanity behaves is in part the result of uncountable and unaccountable economic decisions made by people—singly and in groups—within the context of an increasingly planetary capitalism. "The problem with money," says a character in a recent film, "is that it makes us do things we don't want to do."

Whether or not superhumanity's tendencies are conscious beyond us, individual humans should not be surprised if the aggregate of planetary humanity shows unexpected, emergent, seemingly purposeful behaviors. If brainless bacteria merged into fused protists, that cloned and changed themselves over evolutionary time into civilization, what spectacle will emerge from human beings in global aggregation? To deny the existence of superhumanity by insisting it is merely the sum of human actions is like claiming that a person is merely the sum of the microbes and cells that constitute the body.[8]

Expanding Life

Life today is an autopoietic, photosynthetic phenomenon, planetary in scale. A chemical transmutation of sunlight, it exuberantly tries to spread, to outgrow itself. Yet by reproducing, it maintains itself and its past even as it grows. Life transforms

77
Biology merging with technology. The trans-Alaskan pipeline, snaking across the countryside from Prudhoue Bay to the port of Valdez in Alaska. Graphic example of the "plumbing" of the external oil-using, circulatory system of "superhumanity"—globally connected, industrially contoured, goods and services trading, telecommunicating humanity taken as a global whole. Vladimir Vernadsky in his speculation on the biosphere posited the development of a global layer of thinking and technology called the noosphere. This photograph represents one of the developing conduits of Vernadsky's noosphere.

to meet the contingencies of its changing environment and in doing so changes that environment. By degrees the environment becomes absorbed into the processes of life, becomes less a static, inanimate backdrop and more and more like a house, nest, or shell—that is, an involved, constructed part of an organic being. The members of thirty million species interacting at Earth's surface continue to change the world.

Coming to understand life afresh, we find that species of organisms diverge into new kinds, yet earlier patterns never entirely disappear. Old life forms, the bacteria that run the planet's ecology, are supplemented but not replaced. Although every distinct variety of nucleated being—every species of plant, animal, protoctist, and fungus—perishes, similar new taxa evolve from them or from their kind. Meanwhile, the underlying bacteria march symbiogenetically on.

We find that nature is not always "cutthroat," or, as the poet Alfred Tennyson put it, "red in tooth and claw." Living beings are amoral and opportunistic, as befits their needs for water, carbon, hydrogen, and the rest. They are fractally repeating structures of matter, energy, and information, with a very long history. But they are no more inherently bloodthirsty, competitive, and carnivorous than they are peaceful, cooperative, and languid. Lord Tennyson might just as well have cast nature as "green in stem and leaf."

Among the most successful—that is, abundant—living beings on the planet are ones that have teamed up. Moving inside (or perhaps forcibly dragged inside) another cell, the cyanobacteria that became chloroplasts in protoctist and plant cells weren't lost; they were transformed. So too with the mitochondria—once respiring bacteria—that give your finger muscles the energy to turn this page. Former bacteria, as themselves or parts of larger cells, are still the most abundant forms of life on the planet.

The strength of symbiosis as an evolutionary force undermines the prevalent notion of individuality as something fixed, something secure and sacred. A human being in particular is not single, but a composite. Each of us provides a fine environment for bacteria, fungi, roundworms, mites, and others that live in and on us. Our gut is packed with enteric bacteria and yeasts that manufacture vitamins for us and help metabolize our food. The pushy microbes of our gums resemble department store customers before a holiday. Our mitochondria-laden cells evolved from a merger of fermenting and respiring bacteria. Perhaps spirochetes, symbiotically faded to the edge of detectability, continue to squirm as the undulipodia of our fallopian tubes or sperm tails. Their remnants may move in subtle ways as our microtubule-packed brains grow. "Our" bodies are actually joint property of the descendants of diverse ancestors.

Individuality is not stuck at any one level, be it that of our own species or pondwater *Amoeba proteus*. Most of our dry weight is bacteria, yet as citizens swarming in crowded streets and office buildings, viewing television, traveling in cars, and communicating by cellular and facsimile phone, humans disappear in a global swirl of activity, overwhelmed by emergent structures and abilities that could never be accomplished by individuals or even tribes of human predecessors. No single human can speak to another human, in real time, thousands of miles away. No single human could stand on the moon. These are emergent abilities of superhumanity. Our global activities bring to mind the social insects, except that our "hive" is nearly the entire biosphere. [PLATE 78]

Inextricably embedded in the biosphere, this superhuman society is not independent. At its

78

Queen honey bee surrounded by female workers licking and touching her with their antennae. They are attracted by a chemical called a pheromone which they will distribute throughout the nest subsequently in a buzz of relative hyperactivity. Social organization acts at many levels and magnitudes in stages throughout nature's holarchy. Organisms become societies become superorganisms become individuals super-ordinating into societies again at higher scales.

greatest extent life on Earth—fauna, flora, and microbiota—is a single, gas-entrenched, ocean-connected planetary system, the largest organic being in the solar system. The upper mantle, crust, hydrosphere, and atmosphere of Earth remain in an organized state very different from that on the surfaces of our neighboring planets. Photosynthesis, respiration, fermentation, biomineralization, population expansion, seed germination, stampedes, bird migration, mining, transportation, and industry move and alter matter on a global scale. Life dramatically impacts the environment by producing and storing skeletons and shells of calcium phosphate and carbonate, by caching plant remains as coal and algae residues as oil. Great layers of minerals—sulfides of iron, lead, zinc, silver, and gold—remain in place where they were precipitated by hydrogen gas-producing bacteria.

Minerals not normally associated with life—aragonite, barite, calcite, francolite, fluorite—are produced as crystals and skeletons inside, and exoskeletons or shells outside, living organisms. Plants and microbes induce the formation of "inanimate" substances such as nickel, iron oxides, galena (lead sulfide), and pyrite (iron sulfide, or fool's gold). Humanity's cultures are ranked from a stone age through an iron to a bronze age. Some argue that with the advent of computers Earth has entered a "silicon age." But metallurgy preceded us: human metalworking followed the bacterial use of magnetite for internal compasses by 3,000 million years. *Pedomicrobium*, a soil bacterium found fossilized in gold samples, is thought to precipitate gold ions and thus accumulate gold particles in its sheath. Next to millions of cubic kilometers of tropical reefs built by coral and entire cliffs of chalk precipitated by foraminifera and coccolithophorids, human technology does not seem uniquely grand.

Our destiny is joined to that of other species.

When our lives touch those of different kingdoms—flowering and fruiting plants, recycling and sometimes hallucinogenic fungi, livestock and pet animals, healthful and weather-changing microbes—we most feel what it means to be alive. Survival seems always to require more networking, more interaction with members of other species, which integrates us further into global physiology. Despite the apocalyptic tone of some environmentalists, our species is on its way to becoming better integrated into global functioning. The same technology that now poisons other organisms, stultifies their growth, and does the same to us, may usher in the next major change in biospheric organization.

As anaerobic microbes teamed up to make the swimming ancestors of protist colonists, as mastigotes ingested but did not digest the mitochondria that made them able to invade oxygen-rich niches at Earth's surface, or as fungi and algae combined into lichens that colonized bare rock of dry land, so too the transport of life to new planetary bodies will require teamwork. Astronautics, computers, genetics, biospherics, telecommunications, and other forms of photosynthesis-based, human-sired technology will have to combine with the predecessor technology of other planetmates. The ultimate explosion of life onto its next frontier—that of space—will require the new technology of life itself. The vivification and terraformation, the coming to life of other planetary bodies is not only a human process. Some day recycling ecosystems inside spacecraft may feed humans voyaging to other planets. If humans are to reside in space or voyage beyond Earth's orbit, the plants that feed them, the bacteria that digest for them, the fungi and other microbes that recycle their wastes, and the technology that supports them will surely be along for the ride. The extension of our local thermodynamic disequilibrium into space necessarily will involve representatives of all five kingdoms that have forged new ecosystems, able to transfer energy and cycle matter in isolation from the mother planet, the original functioning biosphere of Earth.

The distinction between space colonization by machines alone or by life with machines mirrors a New Zealand newspaper debate Samuel Butler had with himself. Beginning in 1862 Butler contributed an anonymous article to *The Press* of Christchurch, New Zealand. At the time, he was sheep farming in the Upper Rangitata district of Canterbury Province. Entitled "Darwin on the Origin of Species, a Dialogue," the unsigned article generated spirited protest. Butler joined in, criticizing himself as well as others. Signing his divided opinions under different names, Butler ultimately argued for two diametrically opposed interpretations of machines.

In "Darwin among the Machines," a letter signed by one Cellarius and appearing in *The Press* on 13 June 1863, Butler held that machines were the latest form of life on Earth, posed to take over and enslave their human masters; the rate of evolution and reproduction of machines was prodigious, and without "war to the death" at once it would be too late to resist their world dominion. Then, in a 29 July 1865 article entitled "Lucubratio Ebria," Butler countered Cellarius by saying that human beings were not even human without clothes, tools, and other mechanical accessories. Machines were not a threat to human life, but its indissociable natural extension.[9]

If space vehicles do cut free of human influence, voyaging starry skies as they reproduce on their own, then Cellarius and other Luddites will be vindicated. If, however, machines in space flourish not alone but as intelligent enclosures for a wide variety of other life forms, then the author of "Lucubratio Ebria" will be proven correct.

We place our bets on the latter. Machines, we believe, will flourish in a tightly meshed interface with life—not only human life but a rich assortment of starlight-using life forms. People are essential for making the export of life into the night fantastic even possible. But, like the sperm tail which breaks off once the genetic message enters the egg, so human beings are ultimately expendable. Even without us, a hundred million more years of sun-driven planetary exuberance should be enough to get life off the Earth, star-bound. Other technological species might evolve. Besides, not only humans have begun space exploration.

Setting foot on the moon, Neil Armstrong proclaimed, "One small step for man, one giant leap for mankind." True, in a sense; but he overlooked vast numbers of bacteria on his skin and in his intestine that stepped with him. Life has been expansionist from the beginning. Once it gets a firm toehold in space it may kick off its human shoes and run wild.

Rhythms and Cycles

We and many other animals sleep and wake in cycles that repeat every twenty-four hours. Some ocean protists, dinomastigotes, luminesce when dusk comes, ceasing two hours later. So hooked are they into the cosmic rhythm of Earth that even back in the laboratory, away from the sea, they know the sun has set. Many similar examples abound because living matter is not an island but part of the cosmic matter around it, dancing to the beat of the universe.

Life is a material phenomenon so finely tuned and nuanced to its cosmic domicile that the relatively minor shift of angle and temperature change as the tilted Earth moves in its course around the sun is enough to alter life's mood, to bring on or silence the song of bird, bullfrog, cricket, and cicada. But the steady background beat of Earth turning and orbiting in its cosmic environment provides more than a metronome for daily and seasonal lives.

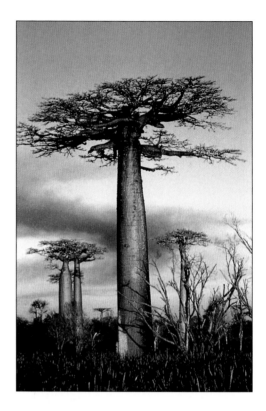

Larger rhythms, more difficult to discern, can also be heard.

Many types of life form encapsulating structures that protect them from temporary dangers of the environment. Propagules of a wide variety of types are miniaturized, viable representatives of mature organic beings. They range from bacterial and fungal spores to protoctist cysts; from plant spores, pollen, seeds, and fruit to the dry eggs of some crustaceans, insects, and reptiles. As such propagules proliferate, natural selection deals with them severely: many die or fail to grow.

Desiccation- and radiation-resistant, most propagules metabolize at an exceedingly slow rate. Spores of bacteria may lay in wait for a hundred years—until rain comes, or phosphorus abounds, or conditions otherwise become less dry and more permissive. Without any dormant seeds or resistant spores, humans survive extraordinary environmental hazards. Houses, clothes, railroads, and automobiles have made possible our expansion from the subtropical home to colder climes. Analogous to spores, cysts, and seeds, these protective structures protect us from harsh conditions. [PLATES 79 and 80]

Recycling greenhouses are enclosed dwellings that contain representative collections of Earth's life. They detoxify poisons and transform wastes into food and back again. One, designed by Santiago Calatrava, will span the entire roof of the Cathedral of Saint John the Divine in New York City. Such "artificial biospheres" miniaturize crucial processes of the autopoiesis of the global ecosystem.

The global ecosystem is not an ordinary organic being. The global system, like all living beings, is energetically open: solar radiation comes steadily in, dissipated heat moves steadily out. But unlike other beings, the global system is closed to material exchange. Apart from the occasional incoming meteor or comet, nothing enters. Apart from the

79 and 80
Progagules of *Adansonia*, baobob trees from the island of Madagascar. Their huge water-borne propagules are seeds. The appearance and evolution of durable desiccation resistant structures appears "fractally" at many levels from bacterial spores to protoctist cysts to the nuts of fruit trees to closed ecological systems proximately built by human beings but ultimately strange new fruit of the biosphere.

4 · Masters of the Biosphere
(pages 68-89)

EPIGRAPHS: The first quotation is drawn from p. 1 of Sorin Sonea and Maurice Panisset, 1983, *A New Bacteriology*, Boston: Jones and Bartlett. The Bassi quotation is cited on p. 221 of Gordon Rattray Taylor, 1963, *The Science of Life: A Picture History of Biology*, New York: McGraw-Hill.

1. For discussion of the microfossils of Barghoorn and of Walsh and other revealers of prePhanerozoic stromatolite and microbial wisdom, see chapters 5 and 6 in Lynn Margulis, 1993, *Symbiosis in Cell Evolution*, 2nd edition, New York: Freeman.

2. Bruno quotation is cited on p. 145 of Wolfgang E. Krumbein and Betsey Dyer, "This Planet is Alive: Weathering and Biology, a Multi-Faceted Problem," in Drever, J. I., 1985. *The Chemistry of Weathering*, Dordrecht and Boston: D. Reidel.

5 · Permanent Mergers (pages 90-117)

EPIGRAPHS: The quotation by Anton van Leeuwenhoek, 1681, is cited in C. Dobell, 1958, *Antony van Leeuwenhoek and His "Little Animals,"* New York: Russel & Russel. The quotation by Charles Darwin comes from p. 204 of his 1868 *The Variation of Animals and Plants Under Domestication,* vol. 2, New York: Organe Judd. The quotation by Stephen Jay Gould comes from his "Foreword to First Edition" of Lynn Margulis and Karlene V. Schwartz, 1988, *Five Kingdoms: An Illustrated Guide to the Phyla of Life on Earth,* second edition. New York: Freeman. The quotation by Lewis Thomas appears on pp. ix-x of Lynn Margulis, Heather I. McKhann, and Lorraine Olendzenski, 1993, *Illustrated Glossary of Protoctista.* Boston: Jones & Bartlett.

1. Hogg, John, 1861. "On the Distinctions of a Plant and an Animal, and on a Fourth Kingdom of Nature." *Edinburgh New Philosophical Journal* 12: 216-225.

2. Haeckel, Ernst. 1889. *History of Creation,* vol 2, first American edition. New York: D. Appleton, p. 45.

3. Haeckel, Ernst. 1887. *Evolution of Man,* vol 1. New York: D. Appleton, p. 180.

4. Smith, David C. 1979. "From Extracellular to Intracellular: The Establishment of a Symbiosis." *Proceedings of the Royal Society, London* 204: 115-130.

5. Wallin, I. E. 1927. *Symbionticism and the Origin of Species.* Baltimore: Williams & Wilkins, p. 8.

6 · The Amazing Animals
(pages 118-139)

EPIGRAPHS: The quotation by Charles Darwin comes from "Difficulties on Theory" in *The Origin of Species,* first edition (1985 reprint). New York: Penguin Books, p. 205. The Shakespeare quotation is drawn from *The Tempest,* Act I, scene two. The quotation by Ralph Waldo Emerson is drawn from his poem *May-Day.*

1. Griffin, Donald R. 1992. *Animal Minds.* Chicago: University of Chicago Press, p. 253.

2. Fischer, A. G. 1965. "Fossils, Early Life, and Atmospheric History," *Proceedings National Academy of Sciences* 53: 1205-1215.

3. Whittington, Harry B. 1985. *The Burgess Shale.* New Haven, Connecticut: Yale University Press, p. 130.

4. McMenamin, Mark and Dianna Schulte McMenamin. 1994. *Hypersea: Life on Land.* New York: Columbia University Press.

7 · Flesh of the Earth (pages 140-157)

EPIGRAPHS: The quotation by R. Gordon Wasson comes from p. 75 of his 1986 *Persephone's Quest: Entheogens and the Origins of Religion,* New Haven, Connecticut: Yale University Press. (Entheogen is Wasson's term for hallucinogens.) The quotation by Franciscus Marius Grapaldus comes from bk. II, chap. 3 of his 1492 *De Partibus Aedium.*

1. Takami, Jun. Cited by Andrey V. Lapo, 1987, *Traces of Bygone Biospheres,* Oracle, Arizona: Synergetic Press, p. 129.

2. Veillard, S. Cited *ibid.,* p. 178.

3. Brasier, Clive. 1992. "A Champion Thallus." *Nature* 356: 382.

4. Blanchot, Maurice. 1981. "Literature and the right to death." Page 46 in *The Gaze of Orpheus and Other Literary Essays* (Lydia Davis, tr.). P. Adam Sydney (ed.). Barrytown, New York: Station Hill Press.

8 · The Transmutation of Sunlight
(pages 158-175)

EPIGRAPHS: The quotation by Georges Bataille comes from pp. 12-13 of his 1988 *The Accursed Share* (Robert Hurley, tr.), New York: Zone Books. Italics are in the original. The quotation by William Blake comes from Blake, William. 1958. *Songs of Innocence and of Experience: Shewing the Two Contrary States of the Human Soul.* Slough: University Tutorial Press of London, pp. 21-22. The quotation by Vladimir Vernadsky comes from Vernadsky, Vladimir I. 1986. *The Biosphere.* Oracle, Arizona: Synergetic Press, p. 11.

1. Went, Frits W. 1963. *The Plants.* New York: Time Inc., p. 16.

2. For a comprehensive view of life in the Proterozoic eon see Bengston, Stefan (ed.). 1995. *Early Life on Earth.* New York: Columbia University Press.

9 · Sentient Symphony (pages 176-197)

EPIGRAPHS: The quotation by Erasmus Darwin comes from his 1794 *Zoonomia; or, the Laws of Organic Life.* The Parmenides quote is a famous fragment of preSocratic philosophy; for alternative translations see Tarán, Leonardo, 1965. *Parmenides: A Text with Translation, Commentary, and Critical Essays.* Princeton, New Jersey: Princeton University Press, pp. xx-xxi. The quotation by Blaise Pascal comes from his *Pensées.*

1. Folsome, Clair. 1985. "Microbes." Pages 51-56 in Tango Parish Snyder (ed.), *The Biosphere Catalogue,* Oracle, Arizona: Synergetic Press.

2. Culver, Stephen J. 1993. "Foraminifera." In Lipps, Jere H. (ed.) *Fossil Prokaryotes and Protists.* Boston: Blackwell Scientific Publications, p.224.

3. Koshland, D. E., Jr. 1992. "A Response-Regulated Model in a Simple Sensory System." *Science* 196: 1055-1063.

4. The most comprehensive source for these little-known events comes from Jones, Henry Festing, 1919. "The Butler-Darwin Quarrel," Appendix C in *Samuel Butler, Author of Erewhon (1835-1902), A Memoir,* vol. II. London: Macmillan and Co., Limited, pp. 446-467. The "mental anachronism" quote can be found on p. 447, the "it never occurred to me" on p. 448, and "how the accident arose" on p. 453. Butler's fascinating but dificult-to-find "evolution books"—*Life and Habit; Evolution, Old and New;* and *Luck, or Cunning*—comprise volumes 4, 5, and 8, respectively, of *The Shrewsbury Edition of the Works of Samuel Butler,* 1924. Henry Festing Jones and A.T. Bartholomew (eds.). New York: E.P. Dutton and Co.

5. Bohr, Niels. 1958. *Physical Science and the Problem of Life.* New York: Wiley, p. 100.

6. Butler, Samuel. 1967. "The Deadlock in Darwinism." in R. A. Streatfeild (ed.), *The Humour of Homer and Other Essays.* Freeport, New York: Books for Libraries Press, pp. 253-254.

7. Calvin, William. 1989. *The Cerebral Symphony.* New York: Bantam. See also Gerald Edelman, 1987, *Neural Darwinism.* New York: Basic Books.

8. For more on the emergence of a superhumanity see Gregory Stock, 1993, *Metaman: Humans, Machines, and the Birth of a Global Superorganism,* London: Bantam. For the relationship between increasingly lifelike machines and increasingly engineered life, between, as the author says, the "born and the made," see *Out of Control: The Rise of Neo-Biological Civilization* by Kevin Kelly, 1994, Reading, Massachusetts: Addison-Wesley.

9. Butler, Samuel. 1863, 1865. "Darwin among the Machines" and "Lucubratio Ebria." Reprinted on pp. 46-53 in Henry Festing Jones (ed.), 1917, *The Note-Books of Samuel Butler, Author of "Erewhon,"* New York: Dutton, pp. 46-53.

Epilogue (pages 198-199)

EPIGRAPH: The quotation by Emily Dickinson is drawn from her letter to Mrs. Holland (c. 1881). Cited in *The Life of Emily Dickinson,* Richard B. Sewall. 1980. New York: Farrar Straus and Giroux, p. 624.

Acknowledgements

The revival of Schrödinger's inquiry into "What is life?" was Peter N. Nevraumont's idea. We are grateful to him for every aspect of this project from the title to the finished book. His bibliophilic attentiveness, so rare in the mercantile world of book publishing, was the motor behind completion of this work.

Michael Dolan collected illustrations and critical comments, prepared the time line, checked dates and references and made numerous other invaluable contributions to manuscript preparation. Connie Barlow edited and José Conde designed at their usual extraordinary levels of competence. We thank Curt Staeger of Paul Smith's College of Arts and Sciences in New York for sending us the tape of his radio show on the subject, and his student Deborah Smith for her statistical survey ("I wanted to research this topic because I want to know what life is. The way I gathered my information was by writing to scientists… From the information I have collected, it seemed like no one really knew what life is. Rather, they chose to tell me the physical features of life, or offered analogies and metaphors. Most students answered by singling out humans specifically.") Thanks to David Abram for his never-say-die animism and his close reading of the manuscript, to Connie Barlow for many suggestions including her lead to Rod Swenson's missing link between thermodynamics and mind. We are indebted to Osborn Segerberg Jr. for sharing his research and experiences. (The secretary of James Watson, co-discoverer of DNA's structure, replied with unintentionally dry humor to Segerberg: "I am so very sorry, but Dr. Watson is unable to answer your question, `What is life?' With our summer meetings now in full swing, his life is most hectic, and he doesn't have time to spare for anything else.") We are grateful to David Bermudes for his emphasis on RNA over exotic cuisine, to Oona West for home-ground coffee and foraminifera facts-on-call, to Ann J. Perrini for her expert management of Nevraumont Publishing Co., to Seth! Leary, Bob Hanie, and his students for excitement and hospitality, to Landi and Rufus Stone for companionship and conversation, to Greg Hinkle for the phrase "symbiosis seen from space," to Tobi Delbruck for manuscript review and Caltech skepticism, and to Emil Ansarov for suggesting the need to examine life's shadow, death. We are especially grateful to Bermudes, Mary Catherine Bateson, Ann Druyan, Freeman Dyson, Ricardo Guerrero, Donna Haraway, Bryce Kendrick, Steven Rose, Michael Rothschild, Jan Sapp, Carl Sagan and E.O. Wilson all of whom made excellent suggestions for improvement.

We thank the great minds who have dug before us and left us their shovels and would, in accordance with our post-structuralist notion of a generalized plagiarism, no doubt credit them completely were it not for our craving for self-aggrandizement, probably itself an evolutionary legacy. Hans Jonas's brilliant history of materialism, *The Phenomenon of Life*, for example, profoundly influenced us. We are very thankful for Eileen Crist's early review of the Butler manuscripts and for Carl Sagan and Ann Druyan's last-minute criticisms and encouragement. We acknowledge the grace of Christie Lyons' illustrations which were based on photomicrographs of organisms taken by David G. Chase, William Ormerod, Lorraine Olendzenski and published others. The rhinoceros beetle came on loan from the department of entomology at the University of Massachusetts, Amherst. We thank so many of our colleagues, acknowledged on page 203 (Sources of Illustrations) who generously provided illustrations. Fern Serna (Newell Color Lab) and Steve Gerard (Photo Researchers) both went the extra mile to help us obtain the best images. We greatly appreciate Donna Reppard, Karen Harney, Thomas Teal, Linda Ward and Lorraine Olendzenski who aided in the innumerable tasks required for manuscript completion. Without the rapid excellent typing of Karen Nelson we never could have finished on time and we are grateful for the assistance of Victor Cavelli, a gentleman printer. Unprecedented financial support for the research on this project came from the Richard Lounsbery Foundation, New York City. The Life Sciences Division of NASA and Dean Frederick L. Byron of the College of Natural Sciences and Mathematics at the University of Massachusetts, Amherst also came to our financial aid. We thank Carolyn Reidy, Dominick Anfuso, Patricia Leasure, Linda Cunningham and William Rosen (Simon & Schuster) as well as Anthony Cheetham, Michael Dover and Lucas Dietrich (Weidenfeld and Nicoloson) for their enthusiastic support of this book.

Any comment we might inspire from you the reader should be directed to us at the New York address of Peter N. Nevraumont, our publisher: Nevraumont Publishing Company, 16 East 23rd Street, New York, New York 10010. Your written response to our prose or illustration is warmly welcomed as the "what-is-life?"-dialogue continues.

Lynn Margulis and Dorion Sagan
Spring, 1995

Sources of Illustrations

[**Plates**] **1.** NASA **2.** CNRI / Science Photo Library **3.** UPI / Bettman Archives **4.** Peter F. Allport **5.** Prof. Woodruff T. Sullivan III, University of Washington, Seattle **6.** NASA **7.** Dr. Johannes H.P. Hackstein, University of Nijmegen, The Netherlands **8.** Prof. Mary Beth Saffo, Arizona State University, Tempe **9.** Nancy Sefton/Science Photo Library **10.** Prof. Bernard Vandermeersch, Université de Bordeaux I **11.** NASA **12.** NASA/Science Photo Library **13.** Prof. Peter Westbroek, University of Leiden, The Netherlands **14.** Dr. Patrick Holligan, Marine Biological Association, Plymouth, England **15.** Prof. David Deamer, University of California, Santa Cruz **16.** NASA **17.** Prof. James A. Shapiro, University of Chicago **18.** Prof. Arthur Winfree, University of Arizona, Phoenix **19.** Prof. Oscar Miller/Science Photo Library **20.** Hans Reichenbach, Braunschweig, Germany **21.** Dr. L. Card/Science Photo Library **22.** Prof. Euan Nisbet, Royal Holloway and Bedford New College, University of London **23.** Prof. Dennis Searcy, University of Massachusetts **24.** Prof. Norbert Pfennig, University Konstanz, Germany **25.** Prof. Andrew Knoll, Harvard University **26.** Prof. Stjepko Golubic, Boston University **27.** Prof. Lynn Margulis, University of Massachusetts **28.** Elso S. Barghoorn (deceased) **29.** (top) Prof. Stanley Awramik, University of California, Santa Barbara; (bottom) Prof. Peter Westbroek **30.** Christie Lyons **31.** Prof. Tore Lindholm, Åbo Akademi, Finland **32.** David Chase (deceased) **33. - 34.** Christie Lyons **35.** Prof. Lynn Margulis **36.** Dr. Heinz Stolp, Bayreuth, Germany **37.** François Gohier/Science Photo Library **38.** Brian Duval, University of Massachusetts **39.** Walker/Photo Researchers **40.** Walker/Science Photo Library **41.** Prof. David John, Oral Roberts Medical School **42.** Prof. Frank E. Round, University of Bristol, England **43.** Christie Lyons **44.** Prof. Charles Cuttress **45.** Prof. Vivian Budnik, University of Massachusetts **46.** EM by Prof. Richard Linck, University of Minnesota; false color by David Gray, Woods Hole Oceanographic Institution **47. - 48.** Prof. Gonzalo Vidal, Uppsala University **49. -50.** Prof. J. Woodland Hastings, Harvard University **51.** R.O. Schuster **52.** Christie Lyons **53.** Dr. Johan Bruhn, University of Missouri **54.** Jukka Vauras **55.** William Ormerod **56.** James G. Schaadt **57. - 58.** Prof. Vernon Ahmadjian, Clark University **59.** Prof. Reinhard Agerer, Universität München, Germany **60.** Prof. Lynn Margulis **61.** Prof. Barbara Thorne, University of Maryland **62. - 64.** Anabel Lopez, Autonomous University of Madrid **65.** Biophoto Assoc./Photo Researchers **66.** Connie Barlow **67.** Prof. Jeremy Pickett Heaps, University of Melbourne, Australia **68.** Michael Dolan, University of Massachusetts **69.** Christie Lyons **70.** Brian Duval, University of Massachusetts **71.** Jim Frazier, from *The Flowering of Gondwana* by Mary White **72.** Prof. David Mulcahy, University of Massachusetts **73.** William Ormerod **74.** Prof. James A. Shapiro, University of Chicago **75.** NASA **76.** Prof. John F. Stolz, Duquesne University, Pittsburgh **77.** Bryan and Cherry Alexander **78.** Kenneth Lorenzen **79. - 80.** Prof. David Baum, Harvard University [**Charts**] **A.** Jeremy Sagan, Cornell University **B.** Charles Keeling, University of Hawaii, and Lloyd Simpson, Center for the Study of the Environment, Santa Barbara **C.** Prof. Lynn Margulis

Index

Book Design

Although there is no direct borrowing, I have been inspired by early scientific books, particularly those printed by Erhard Ratdolt (see Euclid's *Geometriae elementa*, 1482). I have tried to combine the typographic clarity and beauty of these works with a more contemporary sensibility. The text typeface, Minion, designed in 1990 by Robert Slimbach for Adobe Corporation, is loosely based on old style roman typefaces from the late fifteenth and sixteenth centuries. Likewise, the display typeface, Syntax, designed in 1968 by Hans Eduard Meier, is humanistic in structure.

– José Conde / Studio Pepin / Tokyo, Japan

∎

Illustrations by Christie Lyons

Created and Produced by
Nevraumont Publishing Co., New York.
President, Ann J. Perrini